Food in a Planetary Emergency

Dora Marinova • Diana Bogueva

Food in a Planetary Emergency

 Springer

Dora Marinova
Curtin University Sustainability Policy
(CUSP) Institute
Curtin University
Perth, WA, Australia

Diana Bogueva
Centre for Advanced Food
Engineering (CAFE)
The University of Sydney
Sydney, NSW, Australia

ISBN 978-981-16-7709-0   ISBN 978-981-16-7707-6   (eBook)
https://doi.org/10.1007/978-981-16-7707-6

This Springer imprint is published by the registered company Springer Nature Singapore Pte Ltd.
The registered company address is: 152 Beach Road, #21-01/04 Gateway East, Singapore 189721,
Singapore

To Freddy, Alex K, Daniel, Alex B, Lily, Matt, Mira, Nikolay, Ana, Iavor, Nadya, Marin and all others who were with us on this journey of seeing the world with our hearts.

Foreword

In 2018, I was invited to write the foreword for "Less Is More: Reducing Meat and Dairy for a Healthier Life and Planet" – a report produced by Greenpeace International which summarised a large body of scientific evidence about our food systems. It has been very clear for more than a decade that we need to reduce demand for livestock products, and this is now a scientifically mainstream view. The science around climate change, food systems, environmental services, soils and other non-renewable natural resources is pointing irrevocably in one direction – the way we feed the human population needs to change to give a chance for life as we know it on this planet to continue. Unless we shift trajectory and stop over-exploiting the animal species on the land and in the oceans, the well-being of our own species will continue to be threatened.

Unfortunately, ordinary people's perceptions and attitudes towards animal-sourced foods do not correspond well with the science; the public at large is at odds with the scientific evidence. Irrespective of the reasons, changes in food production are happening. This book has set out to examine areas where we are on the transformatory trajectory related to food in an environmental and climate emergency. It builds on the hundreds available studies and meta-analyses about the link between food and environmental impact, including greenhouse gas emissions, and then moves from the global dimensions into innovative industry perspectives, finishing with the contribution that individuals can make through their dietary choices. The last chapter is about Generation Z, whose power emerged with the climate strikes and is already echoing across boardrooms and court cases. How much more evidence do we need before we take action? Or has action already started to occur?

With this book, the two authors show that change is happening, but there is need for this transformative momentum to further speed up. They are not eco-warriors or vegan activists, but two people who have been deeply impacted by the scientific evidence and research into people's attitudes. In my opinion, it will be almost impossible for anybody who reads the skilful synthesis and analysis of the scientific facts laid out in the book to remain neutral or dismiss the calls for a shift.

Every one of us makes a choice with every meal we eat. Food is a basic necessity, but our food choices for a long time have been associated with over-exploitation of livestock, all too often raised in unhealthy, constrained environments. At least 30% of the crops we grow worldwide are used to feed livestock instead of humans. The price for these choices has been not only the spread of non-communicable and zoonotic diseases but also exhaustion of the soils, pollution on the land, waters and air. People need to be creative, innovative and sometimes even look at how things were traditionally done in the past. We also need to re-discover that food gives joy and that eating is social experience which brings together people and planet within the limits of today and the future.

When reading this book, please do not be constrained by the way we have done things until now, but instead look at what we can do from now on, to change our food systems for good.

Fellow of the Royal Society of Edinburgh   Pete Smith
Edinburgh, Scotland
Fellow of the Royal Society
University of Aberdeen, UK
London, UK

Introduction

Much is written about food. The prestigious Gourmand Awards, established in 1995 and considered to be equivalent to the Oscars in the area of food (Gourmand International, 2021), estimate that every year there are about 100,000 books published about food and more than 215 countries compete for the Best in the World title. The book that you are currently reading is also about food. However, it is not a cookbook nor a book arguing about the nutritional and health benefits of particular diets. It is a book about food in a planetary emergency. The issues of food security, shortages and hunger are usually the ones associated with emergency situations. They are not new and we have been dealing with them throughout the entire human history. Relatively more recent are the issues about obesity and overweight, and although these problems seem to prevail in high-income economies, such as Australia, the UK and the USA, they can also be observed in certain sections of the population in poorer countries. There is a lot of medical and public health research that deals with the issues of obesity and overweight. This book is not one of these publications, but it does talk about health – planetary health.

The health of our planet has been compromised, and we talk about food from the perspective of this planetary emergency. It relates to climate change but also to loss of biodiversity, plastics pollution, exhaustion of the planet's soils, freshwater overuse and species' exploitation. The planetary emergency is sounding alarm bells in numerous directions because many of the links between food and the planet have been ignored for a long time. Part of the reason is that starvation persists in many parts of the world and 3.1 million children die annually because of malnutrition (UNICEF, 2013). According to the Chief Executive Officer of the US Fund for UNICEF (2013, para. 3): "The fact that malnutrition is stealing the lives of millions of young children every year shocks the conscience". These needless deaths are avoidable with the current amount of food we produce, but there are many geopolitical and socio-technological barriers that prevent the achievements of zero deaths from starvation across the globe. Again, despite being extremely important, this book is not about these issues.

What we are writing about is the threat that the current production systems are posing to the global ability to produce food. The climate of our planet is changing triggered by anthropogenic activities, and whether we are prepared to admit it or not, our food choices are a major contributor to the current environmental emergency. Energy, transportation, buildings, industrial development and many other economic sectors are significant contributors to the climate emergency, but unless our food production systems and nutritional choices change, humanity is facing a bleak future (Pelletier & Tyedmers, 2010). This book is about how to avoid such despondent projections. We need to face the truth and be aware of the scientific evidence to be able to shift course. In this book, we give many facts and cover new developments, but we also realise that the transformation in our food systems is happening as we write. Innovation is penetrating the food industry, and there is increasing awareness about the scale of the problems and challenges ahead. Speaking out about the climate and environmental emergency in support of the health of the planet takes courage. We are part of a global movement that has already started, and we are inspired by the opportunities.

As of June 2021, 34 countries and 1947 local governments have declared a climate emergency, comprising more than 1 billion people (Cedamia, 2021). The number of local councils in Australia where we live is 102, covering 36% of the country's 25 million population. In Bangladesh – a country which is projected to be significantly impacted by climate change, the National Assembly declared an emergency in November 2019, covering 100% of its 161 million population. The European Parliament also declared a climate emergency in November 2019, which covers 448 million people. Many island economies exposed to the threats of sea-level rise have similarly made national declarations, including Japan, the Maldives, Malta and New Zealand. Climate change is increasingly being recognised as an emergency to which we need to respond urgently with decisive actions within the 5–10 years of opportunity we have to shift the current trajectory of global warming and keep any temperature increases within 2 °C, preferably 1.5 °C, according to the 2015 Paris Accord. A lot of changes are already occurring in the areas related to energy, transport and buildings.

Let's take, for example, the energy sector. The 2021 report by the International Energy Agency (IEA, 2021) is already reflective of a global shift towards Net Zero by 2050, offering a roadmap with milestones that results in immediate and large-scale deployment of all available clean and efficient energy technologies, such as hydro, solar and wind, to trigger a huge decline and potential elimination of fossil fuels from the energy system. The report clearly states that all technologies required for such a transition already exist and their deployment will allow to significantly reduce the energy intensity of the economies as measured through their gross domestic product (GDP) and decrease the share of fossil-fuel based sources used. In addition to not pursuing any new coal, gas or oil projects, the roadmap identifies significant room for innovation and creativity linked to electric vehicles, battery storage of energy, direct air capture and storage, carbon utilisation, and hydrogen production. The shifts that are needed to achieve Net Zero and respond to the climate emergency are significant, but only a small component of them, estimated at 4%, requires behaviour change (IEA, 2021).

However, when it comes to food, behaviour change is a major factor. A similar roadmap for a global shift to Net Zero is also needed for food production. In addition to climate change, we need roadmaps to reduce the transgression of the remaining eight other planetary boundaries (see Fig. 1) as well as other known challenges, such as plastics pollution and soil health. With climate change, we are approaching the zone of uncertainty and increased risk. The situation is the same with land-system change where the conversion of native habitats into areas suitable for human activities, and predominantly agriculture, is a substantial factor.

We have already exceeded two other planetary boundaries, namely in relation to the biochemical flows of phosphorus (P) and nitrogen (N), and biosphere integrity as represented by the species extinction rate within the boundary of genetic biodiversity. This means that we are now operating in high risk beyond the zone of uncertainty. We are wondering how many, if any, risk managers would allow an organisation or corporation to conduct financial or any other business activities in a

Fig. 1 Planetary boundaries. (Source: J. Lokrantz/Azote based on Steffen et al. (2015). Note: *P* phosphorus, *N* nitrogen, *E/MSY* extinctions per million of species per year, *BII* Biodiversity Intactness Index)

zone of high risk and uncertainty. Our university would not allow this, but we are all living as a human species on a planet in a zone of high risk! Who are our risk managers and where are they? Don't they see the planetary emergency situation?! This is what our book is about – it is a wake-up call to everybody irrespective as to what their role or place in society is, irrespective as to where they live. We all need to become risk managers for the health of the planet in this current emergency situation.

To make things even more complex, there are also many unknowns in this planetary emergency (see Fig. 1). Science is yet to identify the functional importance of biodiversity and the consequences from the high rates of species extinction. With the massive scale of land conversion and water-use changes, most extinctions are probably happening without being properly observed or analysed. Driven mainly by human activities, the current rate is estimated at 100 extinctions per million species per year (E/MSY) and is expected to reach 1500 E/MSY by the end of the twenty-first century (Lumen, 2019). We can only guess what this means to us – loss of ecosystem services, loss of beauty and wonderment, and also loss of opportunities for new medicinal drugs.

It is extremely important to acknowledge which human activities have caused and continue to contribute disproportionately higher to the transgression of the planetary boundaries and moving into the risk zone. The disturbance of the planet's phosphorus cycle is entirely driven by agriculture and the use of mineral fertilisers. This varies across regions, with North America, Europe, parts of Asia, and smaller areas in Latin America and Oceania significantly exceeding the safe zone (Steffen et al., 2015). The situation is very similar in relation to the nitrogen cycle with the major anthropogenic perturbation arising again from the application of fertilisers.

The analysis by Campbell et al. (2017) shows that agriculture has played a major role in destabilising the Earth's system and triggering the planetary emergency. It has been the main factor in the transgression of the planetary boundaries and is a significant or major contributor to changes in zones that are currently seen as still safe or approaching increased risk. The complexity of providing food in a planetary emergency is described as a "triple burden" – some do not have enough calories, others do not have the sufficient levels of nutrients, and many more are consuming too many calories (Campbell et al., 2017). By adopting a stance as risk managers, we need to shift away from the simplistic mantra of more production (Ingram, 2017) and look at the complexity of food choices where demand and individual preferences need to transition to more plant-based options.

Food packaging is another big area of concern as plastic waste is now threatening the health of marine, riverine and terrestrial environments across the globe. We need to reconsider not only how we produce food but also how we deliver it and how much is wasted or lost in the process. These are all aspects that we cover in this book.

For the first time in the history of food, we need to deal with a planetary emergency. There will always be books about how to cook and prepare food, but this publication is about how to preserve the ability of the planet to support and feed its human population. It requires changes in people's preferences as well as the adoption of new technologies and novel concepts about food. According to McClements (2019, p. 5), "[a] well-functioning global society should address a hierarchy of needs associated with foods, ranging from meeting people's basic nutritional requirements, ensuring what people eat is safe, maintaining a sustainable environment, providing rewarding employment, and contributing to a healthy food culture". This sounds like and is a complex task, but a food system based on constant extraction and use of non-renewable resources and exploitation of a limited number of species, many of them sentient beings, is bound to fail (Mann, 2021). From "an ecological serial killer" (Harari, 2019), homo sapiens needs to redeem themselves, and a major step on this journey is the choices we all make about food.

The book is structured in three parts in which we move from the global picture to industry issues and to the individual consumer. There are two main threads that go across the entire book. The first is that if the current trends and practices in the consumption of animal-source products, including livestock and fish, continue, we are on a trajectory to make life on this planet very uncomfortable and lose the ability to feed the global human population. This will further widen the existing inequalities and create scary scenarios of competition over resources. The Western-type diet, which for many epitomises progress and development, is in various ways responsible for this. It results in devastating effects for the planet while also negatively impacting human health. A shift to better and smarter food behaviour is needed on the part of consumers and producers. The second thread is that we can find better solutions and ways to produce foods that bring hope for the future. Some are easier and more acceptable than others, but if we put our minds and efforts together, we can start seeing much more optimistic scenarios. This is particularly important for the younger generations who will inherit the planet, and its ability to feed the human species, from us. If we make the most-needed changes, these generations can remain optimistic. We hope that the book has this double-pronged approach of stating the warning and often-alarming facts as they are, and kindling hope in the creativity and power of the human race to break free from the current trajectory and find inspiration for the future.

In the part on global perspectives, we start with the concept of sustainability and link it to a planetary diet that would allow everybody to live a comfortable and healthy life. We then explain the link between climate change and food and also between dietary choices and environmental deterioration. The last chapter in this part talks about the unreasonable practices of wasting food and relying on plastics for packaging. All these chapters are also putting forward some solutions. The chapters can be read on their own as they present a particular facet of the multitude of possible lenses we can apply to investigate the issues about food in a planetary emergency. We do not pretend to cover all possible responses and are aware that a lot of good effort is happening across the globe to change the current trends, but we have attempted to capture some of it.

The next part that looks at the food industry and marketing begins by reconceptualising agriculture as closed-loop activities. Circular agriculture is a shift away from the current exploitative ways of growing food that destroy soil fertility, contribute largely to greenhouse gas emissions and leave little space for wilderness to a model that regenerates and cares for nature. Sustainability transitions are also important in finding innovative ways of farming and producing food. We know that the human body needs proteins, and we have a chapter that specifically looks at proteins that are not livestock-based, such as plant-sourced or produced with the latest technological advances. How we market food and the need for behaviour changes in our eating pattens are also discussed in this part.

Eating is ultimately a way to sustain the human body, and the last part of the book is about the individual food choices we make. We talk about mainstreaming flexitarianism and limiting the intake of animal-based foods by replacing them with better choices, particularly vegetables, legumes, fruits and nuts. What we eat also impacts our health, and a chapter in this part presents an overview of mitigating diseases through our food choices. The final chapter focusses on Generation Z, the world's largest population group of upcoming consumers, and their food choices. Generation Z will decide the future of many of the innovations and transitions we describe in this book. They will also come up with their perspectives not only on food but also about the world we will leave behind.

We have carried out a lot of desktop research in putting together this book. Many thanks go to all authors whose publications, websites and blogs have helped build the picture of global food. This picture contains a range of shades, but the ones we want to keep for the time ahead are the vibrant and promising colours that will make the future bright.

References

Campbell, B. M., Beare, D. J., Bennett, E. M., Hall-Spencer, J. M., Ingram, J. S. I., Jaramillo, F., … Shindell, D. (2017). Agriculture production as a major driver of the Earth system exceeding planetary boundaries. *Ecology and Society, 22*(4), 8. https://doi.org/10.5751/ES-09595-220408

Cedamia. (2021). *Climate emergency declarations.* https://www.cedamia.org/global/

Gourmand International. (2021). *Best in the world.* https://www.cookbookfair.com

Harari, Y. N. (2019). *21 lessons for the 21st century.* Vintage.

Ingram, J. (2017). Perspective: Look beyond production. *Nature, 544,* S17. https://doi.org/10.1038/544S17a

International Energy Agency (IEA). (2021). *Net Zero by 2050: A roadmap for the global energy sector.* https://iea.blob.core.windows.net/assets/4482cac7-edd6-4c03-b6a2-8e79792d16d9/NetZeroby2050-ARoadmapfortheGlobalEnergySector.pdf

Lumen. (2019). *Loss of biodiversity.* Boundless Biology. https://courses.lumenlearning.com/boundless-biology/chapter/the-biodiversity-crisis/

Mann, A. (2021). *Food in a changing climate.* Emerald Publishing.

Marinova, D., & Bogueva, D. (2019). Planetary health and reduction in meat consumption. *Sustainable Earth, 2,* 3. https://doi.org/10.1186/s42055-019-0010-0 or https://rdcu.be/b2Riz

McClements, D. J. (2019). *Future foods: How modern science is transforming the way we eat.* Springer Nature.

Pelletier, N., & Tyedmers, P. (2010). Forecasting potential global environmental costs of livestock production 2000–2050. *Proceedings of the National Academy of Sciences of the United States of America (PNAS), 107*(43), 18371–18374. https://doi.org/10.1073/pnas.1004659107

Steffen, W., Richardson, K., Rockström, J., Cornell, S. E., Fetzer, I., Bennett, E. M., … Sörlin, S. (2015). Planetary boundaries: Guiding human development on a changing planet. *Science, 347*(6223), 1259855. https://doi.org/10.1126/science.1259855

United Nations International Children's Emergency Fund (UNICEF). (2013). *UNICEF: Too many children dying of malnutrition.* https://www.unicefusa.org/press/releases/unicef-too-many-children-dying-malnutrition/8259

Contents

Part III   Individual Perspectives

About the Authors

Dora Marinova is Professor of Sustainability at the Curtin University Sustainability Policy (CUSP) Institute, Australia, which she established in 2007 and where she was director between 2015 and 2018 and deputy director between 2007 and 2015. Dora started as an academic at the University of National and World Economy, Sofia, Bulgaria, and previously worked at Murdoch University, Australia, where she was head of school of the Institute for Sustainability and Technology Policy (ISTP). Her areas of research interest cover sustainability, innovation and sustainometrics.

Dora has 370 refereed publications and has supervised more than 70 PhD students to successful completion. She has published in high-impact journals, including *Scientometrics*, *Journal of Cleaner Production*, *Journal of Econometrics*, *Nanotechnology* and *Sustainability*. Two of the books she co-edited – *Impact of Meat Consumption on Health and Environmental Sustainability* and *Environmental, Health and Business Opportunities in the New Meat Alternatives Market* – received "Best in the World" at the prestigious 22nd and 24th Gourmand Awards, in the categories of Sustainable Food and Vegetarian Food, respectively. Between 2013 and 2020, Dora served as a member of the Centres of Research Excellence Peer Review Panels on Population Health and Health Services, National Health and Medical Research Council (NHMRC), Australian Government. She is an elected member of the Australian Institute of Aboriginal and Torres Strait Islander Studies (AIATSIS) and elected fellow of the Modelling and Simulation Society of Australia and New Zealand (MSSANZ) and the International Environmental Modelling and Software Society (IEMSS) headquartered in Switzerland. For the period 1970–2021, Scopus lists her as Australia's top author and 23rd in the world in the area of sustainability. Four of Dora's papers co-authored with Diana Bogueva are ranked in the top 5% in the world's research output by Altmetric for the worldwide attention in the social media they have received.

Diana Bogueva is a proactive interdisciplinary researcher focused on consumer behaviour change, alternative proteins, future novel food processing technologies, food sustainability and harmonisation. Currently, she is the manager of the Centre for Advanced Food Engineering, University of Sydney, and a research fellow at the

Curtin University Sustainability Policy (CUSP) Institute, Australia. Diana's work has won two international awards: the Australian National Best Book winner in 2019 and the World's Best Book award 2020 in the Vegetarian Food category at the prestigious 24th Gourmand Awards, considered equivalent to the Oscars in the area of food, for her co-edited book with Prof. Marinova *Environmental, Health and Business Opportunities in the New Meat Alternatives Market*. In 2020, she won the Faculty of Humanities Journal Article of the Year Award at Curtin University for her co-authored paper "Planetary health and reduction in meat consumption". Altmetric scored four of her papers in the top 5% of all research output in the world. She is also a finalist in the 10th International Book Award at America's Book Fair 2019 for her co-edited book *Handbook of Research on Social Marketing and Its Influence on Animal Origin Food Product Consumption*.

Diana is an active member, Working Groups Director and Ambassador for Australia at the Global Harmonisation Initiative, headquartered in Austria. She is an Elected Member-in-Large at the Non-thermal Processing Division of the international Institute of Food Technologies headquartered in the USA. Diana is a program producer and presenter at the Australian Special Broadcasting Services (SBS) Radio and in 2017 was a finalist at the New South Wales' Premier Multicultural Media Awards for best radio and audio reports. She is also a member of the International Federation of Journalists headquartered in Belgium.

List of Figures

List of Tables

Part I
Global Perspectives

Chapter 1
Sustainability and a Planetary Diet

Abstract The chapter interprets the meaning of sustainability and the concept of planetary health that links together human and ecological well-being. Food production and consumption need to sit within the planet's geophysical boundaries and a planetary diet should allow the projected 10.9 billion people to live comfortably on Earth. The transitioning to such a better planetary diet with more plant-based options has to happen through a double-pronged approach of innovation and behaviour change. Instead of bleak scenarios for the future, there should be hope and enthusiasm as we can make a change each time when we decide what to eat.

Many years ago, we migrated from Bulgaria to live in Australia – Dora (the first author of this book) moved to Perth, Western Australia while Diana (the second author of the book) moved to Sydney, New South Wales. We left what used to be socialist Bulgaria and fell in love with the blue skies, clean air, beautiful beaches and freedom in Australia. What we did not realise at the time was that with this transition of about 12,000 kilometres between the breaking-down socialist economy and the free market with multitudes of brands and the glitter of packaging, we have also made a major environmental jump. Our ecological footprint in Bulgaria was just above one planet – in other words, if everybody in the world lived the way we did in Sofia, everything the global human population was consuming could have been contained within the biocapacity of 1.1 earths (Global Footprint Network, n.d.). This included modest quantities of meat, reasonable amounts of dairy and large quantities of seasonal fruits and vegetables. In Australia, our ecological footprint overnight became 4.3 earths. To us personally that meant improved accessibility to a variety of foods, better living standards, opportunities for travel and recreation, meaningful and enjoyable work opportunities. At the time of our settling in Australia, we could hardly make sense of the word "sustainability" which did not exist in Bulgarian and the concept was difficult to translate.

Now we have established ourselves as academics who work in the area of sustainability researching and teaching how to balance and integrate environmental, social and economic priorities. We have managed to reduce our individual

D. Marinova, D. Bogueva, *Food in a Planetary Emergency*,
https://doi.org/10.1007/978-981-16-7707-6_1

ecological footprints to 3.3 earths and COVID-19 cut them further down to 1.4 earths by restricting international travel. What simply used to be common sense and normal life in socialist Bulgaria, as well as in many other countries after the end of the Second World War, is now a goal that we personally, Australia and the rest of the world are trying to rediscover, re-invent and achieve – namely, living within the planetary boundaries of planet Earth. In the meantime, the global population numbers have increased – from 5.4 billion in 1991 to 7.8 billion in 2020 (Worldometer, 2021), and this planet has to be shared with more people. Projections are that by 2100 global population will stop growing after having reached 10.9 billion (DESA, 2019). All these people would want to live a good and meaningful life on this planet. Is this possible? Are we looking at other habitats for humans, such as Mars or the Moon, because we won't be able to live sustainably on planet Earth? Or can we see changes that allow the entire human population to exist comfortably on the planet we have?

The Global Sustainable Competitiveness Index Report released in 2020 (SolAbility, 2020) represents the competitiveness of 180 nations based on 127 quantitative indicators which measure the countries' natural capital – availability and level of depletion, resource efficiency – or operational competitiveness in a resource-constrained world, social capital – health, security, freedom, equality and life satisfaction, intellectual and innovation capital – capability to generate wealth and jobs through innovation and value-added industries in globalised markets, and governance performance – investments in infrastructure, market and employment structure, provisions for sustained and sustainable wealth generation. Bulgaria is ranked 34th and ahead of Australia which is in 45th position (SolAbility, 2020). A colleague of ours told us: "You failed!". This was a succinct assessment of the efforts of two persons who have moved from Bulgaria to Australia and were working in the area of sustainability for an extended period of time. Yes, we also felt that we have failed.

Both of us have failed our new adopted homeland not because it performed worse than Bulgaria and 43 other countries, but because all the global population is now in a state of a planetary emergency. It is no longer a matter of whether we outperform each other in a competitive race, but a fight for survival. We all, as a human population, are failing and this book is an attempt to respond to some of these failures by tackling the most basic requirement for survival of any species – food. Can we sustain life on Earth and how should the powerful human species feed themselves?

In this chapter, we start with interpreting the meaning of sustainability and the concept of planetary health which links together human and ecological well-being. Then we discuss how food production and consumption sit within the planet's geophysical boundaries and the need for a planetary diet that can allow the projected 10.9 billion people to live comfortably on Earth. The transitioning to such a diet that respects the boundaries of the planet will need to happen through a double-pronged approach of innovation and behaviour change. Relying solely on one, be it innovation or behaviour change, is unlikely to deliver the shifts that can preserve planetary health into the future. However, the combination between innovation and changes in people's food-related behaviour can result in a positive outlook for the human race by 2100 and hopefully, much earlier before that. Instead of a bleak scenario, we want to build hope and enthusiasm. In this point in time, the future is shaped by what food we consume more than by the energy we use, the cars we drive or the buildings we live or work in. This future starts each time when we decide what to eat. Are we ready?

Defining Sustainability

A description of sustainability is the one mentioned above, namely integrating simultaneously environmental, economic and social priorities. This is in the spirit of the highly quoted 1987 Brundtland's report *Our Common Future* (World Commission on Environment and Development, 1987) and the many other international policy forums that followed, such as the 1992 Earth Summit in Rio de Janeiro, 2002 World Summit on Sustainable Development in Johannesburg, 2005 World Summit on Social Development in New York, 2015 Paris Agreement related to climate change and UN Sustainable Development Goals (SDGs), to name a few. Such a definition is very much anthropocentric and puts the human species and its activities at the centre of the existence of life on Earth. However, even within such a human-centric approach which ignores the existence of a myriad of other species of fauna and flora, sustaining the ability of the planet to produce food should be at the core of our understanding of sustainability.

Although food is a basic necessity of a paramount importance, the majority of policy efforts have seen it only as one element of human existence. This would have been okay, if we had enough resources and space on this planet to produce the food we need whilst maintaining a good quality of life and persevering with economic activities. However, as the story of our own ecological footprint indicates, this is not the case. Despite all progress made in the areas of energy, transport, buildings and industry, Europe's ecological footprint still requires 2.8 planet Earths (WWF EU, 2019). The impact of the COVID-19 pandemic and the imposed lockdowns in China, Italy, UK and the rest of Europe, India and the US State of California, was short-lived with greenhouse gas emissions bouncing back. Overall, there was only a 6.4% decline in 2020 compared to 2019 which is below the 7.6% annual cuts required to contain temperature raises within 1.5 °C higher than pre-industrial times as aspired in the 2015 Paris Agreement (Tollefson, 2021). Food and livestock-based products continue to contribute towards high levels of greenhouse gases, including the more powerful methane and nitrous oxide. The current climate change and environmental emergency declared since 2016 by 15 countries and 1925 places across the globe covering one billion people (Cedamia, 2021), also confirm that we are living on an endangered planet with increased frequencies of extreme weather events and fast expanding biodiversity loss. Sustainability then becomes a meaningless concept if we cannot provide and sustain the ability of the planet to satisfy the human requirements for food. What should be at the core of the concept of sustainability is our food systems. The Harvard University T. H. Chan School of Public

Health (2021, para. 1) similarly states that in the multifaceted nature of sustainability, "the food production system and our diets play a crucial role".

We are already seeing some positive changes with numerous calls for a shift towards making more sustainable choices in food production and consumption (Tirado et al., 2018; WRI, 2018; WWF EU, 2019; Crippa et al., 2021). This equally applies to land and ocean areas. While vast land areas of native vegetation are constantly being converted for the needs of the intensive livestock industry, industrialisation has also spread its fishing and trawling nets over marine environments causing devastation of gigantic proportions. It is not just the land and soil fertility that have been exploited to the brink of desolation, the global oceans are also in crisis from decades of uncontrolled exploitative practices (FAO, 2018; WWF EU, 2019). The majority of the world fish stocks, around 60%, are either fully fished or overfished (FAO, 2018).

Hence, for the purpose of this book, we define sustainability as maintaining the ability of the planet to produce enough quality food for human consumption. This definition has two main aspects. The first relates to planetary health which determines the ability of the planet to produce food, and the second – to how we understand enough quality food. This latter aspect is linked to the types of diets that human population should be adopting and practising. We explore these two aspects in the remainder of the chapter.

Planetary Health and Boundaries

Planetary health emerged as a new area of research and action which calls for protecting nature to safeguard human health (Myers & Frumkin, 2020). In fact, planetary health is the opposite of the current environment and climate emergency. Changing human culture to better embrace planetary consciousness is a necessary condition for sustainability – a condition that is manifested every day with every meal we take. The biophysical boundaries of planet Earth are already stretched to allow nature to rejuvenate and recover (Steffen et al., 2015). Our food production systems are largely responsible for global warming, significant land-use changes and exceeding the planet's boundaries for phosphorus, nitrogen and biodiversity loss. According to the Rockefeller Foundation–Lancet Commission, "human health and the health of our planet are inextricably linked, and… our civilisation depends on human health, flourishing natural systems, and the wise stewardship of natural resources. With natural systems being degraded to an extent unprecedented in human history, both our health and that of our planet are in peril" (UNFCCC, 2021, para. 1).

There is ample evidence that reduction in the consumption of all animal-based products, including cutting down on meat or ruminant meat, is associated with lower greenhouse gas emissions and reduces demand for land (e.g. Hallström et al., 2015). Such a reduction will not only improve the health of the biophysical systems on the planet but also the well-being of human population, particularly in countries with high consumption of animal-based foods and meat in particular. For example, about half of all American adults have one or more preventable chronic diseases, often associated with overweight and obesity (Dietary Guidelines Advisory Committee, 2015). There are numerous co-benefits from shifting to food intake that is lighter in its environmental footprint and better for human health. A world where all species thrive and the Earth's natural systems are healthy is possible to achieve, but only if we make our dietary choices more sustainable.

There are many definitions of a sustainable diet and they all emphasise the ability to produce quality and nutritious food for generations to come. For example, the Global Panel on Agriculture and Food Systems for Nutrition (GPAFSN, 2016, p. 32) describes high-quality diets as "those that eliminate hunger, are safe, reduce all forms of malnutrition, promote health and are produced sustainably i.e. without undermining the environmental basis to generate high-quality diets for future generations". Respect and protection of biodiversity are increasingly becoming part of the vision for sustainable diets (e.g. FAO, 2012) but the link between nature and human health is rarely recognised in dietary guidelines. In 2020, the same Global Panel (GPAFSN, 2020) concluded that the present food systems are no longer fit for purpose sounding loud alarm bells about the spiral of decline in human health and of environmental systems. The Panel developed four goals to better nourish the global population and protect the environment. Although there was no particular explicit emphasis on reducing the intake of animal-based foods, the report showed how more livestock raising leads to malnutrition in all its forms, including hunger and obesity, as well as eutrophication and worse environmental outcomes. Less than a quarter of the 83 national dietary guidelines analysed by Gonzales Fisher and Garnett (2016) include a recommendation about limiting the intake of red meat. Only a few countries, namely Brazil, France, Germany, Holland, Qatar, Sweden and USA, make a reference to environmental factors in their dietary guidelines while numerous attempts to make this link explicit by environmentalists, including organisations such as Doctors for the Environment, have proven unsuccessful in Australia. The World Health Organisation (2020) also does not make the connection between the natural environment and humans' food intake in its healthy diet guidelines. It was not until 2019 that the new concept of a planetary health diet emerged which brought together health and planetary boundaries.

Planetary Health Diet

In 2019, the EAT-Lancet Commission, comprising 37 leading scientists from 16 countries, produced a report (Willett et al., 2019) which analysed the health and environmental boundaries for possible diets that can generate a safe operating space for food systems with a human population reaching 10 billion within the biophysical constraints of the planet. These dietary choices referred to as "planetary health diet" combine scientific health-related and environmental targets for an intake of 2500 kilocalories per day.

Table 1.1 Planetary diet recommendations

Food group	Grams (possible range) per day	Caloric intake (kcal/day)
Whole grains (e.g. rice, wheat and corn)	232	811
Starchy vegetables (e.g. potatoes and cassava)	50 (0–100)	39
Vegetables (e.g. dark green, orange and red vegetables)	300 (200–600)	78
Fruits (e.g. apples, oranges and berries)	200 (100–300)	126
Dairy foods (e.g. whole milk and cheese)	250 (0–500)	153
Protein sources (exchangeable)		
Red meat (e.g. beef, lamb and pork)	14 (0–28)	30
Poultry (e.g. chicken and turkey)	29 (0–58)	62
Eggs	13 (0–25)	19
Fish	28 (0–100)	40
Legumes (e.g. dry beans, lentils, peas and peanuts)	75 (0–175)	314
Soy foods (e.g. tofu, tempeh and edamame)	25 (0–50)	112
Tree nuts	25	149
Added fats		
Palm oil	6.8 (0–6.8)	60
Unsaturated oils (e.g. olive, sunflower, rapeseed and walnut oil)	40 (20–80)	354
Lard or tallow	5 (0–5)	36
Added sugar		
All sweeteners (e.g. sugar, corn and maple syrup, stevia and xylitol)	31 (0–31)	120
TOTAL		**2503**

Source of data: Willett et al. (2019)

Table 1.1 summarises the food groups and intakes from such a planetary diet. Although the planetary diet allows for modest intake of animal-based products, its focus is on plant-based foods. It allows enough flexibility and individual preferences, including cultural traditions. At its core is a flexitarian approach to food and we discuss this in a separate chapter later in the book. A typical plate of those following a planetary health diet would be half-full with vegetables and fruits, a third will be whole grains, followed by plant-sourced proteins and smaller amounts of unsaturated oils, some added sugar and starchy vegetables. If consumed at all, animal protein and dairy would occupy a very small section of the plate. In many ways, this diet resembles the choices of foods we used to eat in Bulgaria and is still the case in other traditional societies, such as in Bangladesh, Indonesia or Turkey. Although the planetary health diet has attracted some criticism for lacking full consensus across the globe and for different views among nutrition scientists (e.g. Hauver, 2019), its main message is difficult to challenge – that is, we need to reduce the consumption of animal-based foods and transition to better dietary choices if we are to deal successfully with the current planetary emergency.

Transitioning to Better Diets

Put simply, better diets are those that are healthy and come from food systems which allow the natural environment to continue to produce food. There are several approaches in transitioning to better diets. On the one hand is the production system and on the other, the consumer choices. From a production perspective, the two general approaches are to limit further expansion of agricultural land and protect marine environments as well as endorsing innovation and novel ways of growing and producing food. The consumer can contribute towards a shift to better diets by optimising calorie intake and opting for more sustainable plant-based options by restricting the intake of animal based-foods.

Limit Expansion of Agricultural Land and Protect Marine Environments

Numerous calls are being made to restrict the expansion of agricultural land and stop clearing of native vegetation for the purposes of animal grazing and feed production (Tirado et al., 2018; Willett et al., 2019). Between 2015 and 2020, 10 million hectares of native forests have been lost each year with agriculture, primarily cattle ranching and cultivation of soybeans for animal feed, acting as the main driver of deforestation and forest degradation (FAO & UNEP, 2020). Forests are critical for biodiversity, but other areas of native vegetation, such as savannahs and shrubs, are similarly important and conversion for agricultural and livestock purposes triggers losses of ecoservices, including the ability to mitigate climate change. A planetary diet with limited consumption of animal-based foods can put a stop on the continuous expansion of agricultural land and limit the exploitation of the global oceans. The demand for land increases proportionately with the presence of animal foods in one's diet – from 1.08 hectares (more than two football fields) per person per year on the standard American diet to 0.13 hectares for vegans (Peters et al., 2016) – an 8.3 times difference! A study by Oxford University, based on data from 38,000 farms in 119 countries about 40 food products constituting 90% of all food consumed, estimated that we will need 75% less agricultural land if all people in the world went vegan (Poore & Nemecek, 2018). This will also drastically improve the health of the global oceans. In addition to food not generating as much greenhouse gas emissions and pollution, this will free up space on the land and in the oceans for carbon storage through native vegetation and phytoplankton.

Restoring the health of the marine environments through regulating fish harvesting, putting an end on overfishing and the use of destructive practices is one of the targets under SDG14 Life below Water. Furthermore, subsidies to practices which contribute towards overcapacity exploitation and overfishing as well as illegal, unregulated and unreported fishing should also be prohibited. All these measures are supposed to be put in place by 2020; however, progress has been slow

(Milo-Dale, 2020). It is unthinkable of achieving a sustainability global agenda without protecting the marine environments from overfishing as well as from pollution generated from fishing nets estimated to be responsible for at least 46% of the plastic in the Great Pacific Garbage Patch (Lebreton et al., 2018). The situation with SDG15 Life on Land is similarly bleak with agriculture and livestock in particular causing decline in the areas covered with forests, degradation of soil and water resources, and driving 27% of the Red List's threatened species to extinction (DESA, n.d.). Removing subsidies favouring livestock farming is also long overdue (Shapiro, 2016).

The safeguarding of the oceanic environments has shifted from marine protected areas which are often subject to political manipulation and biases, to key biodiversity areas whose health is essential for threatened species and aquatic populations of high density (Edgar et al., 2008). On the land, protection is put in place through areas of limited or controlled public access, such as wildlife sanctuaries, national parks and nature reserves. As of 2020, only 46% of marine and 44% of terrestrial key biodiversity areas were protected (UNEP-WCMC, IUCN & NGS, 2021). Overall, only 19% of all key biodiversity areas across terrestrial, marine and freshwater systems were completely covered as protected spaces while 39% had no protection at all (UNEP-WCMC, IUCN & NGS, 2021). These protected areas represent only 7.3% of all oceans and 14.9% of all land surface (Marris, 2019) with the remaining surfaces being left to be exploited with very limited regulation and control. One gloomy result from this lack of consideration for the other species inhabiting the planet is that in today's world only 4% of all mammals on Earth by weight are wild animals and the other 96% are livestock (Marris, 2019).

Many conservationist organisations are calling for an international goal of making 30% of oceans and land protected for nature by 2030 (IUCN, 2021). This can only be achieved with a drastic transition in human diets away from animal-based foods. The choice is clear – we either protect nature and wilderness or face continuous species extinction. Around 20% of native species in all terrestrial biomes have declined and a million (500,000 plants and animals plus 500,000 insects) of the eight million species on Earth are threatened by extinction (IPBES, 2019). Only a transition to better dietary choices can put a stop on this unprecedented deterioration of the biological world at the expense of people's gluttony for animal proteins.

Endorse Innovation and New Ways of Growing Food

Economic development and adoption of technologies through human history seem to have followed distinctive trends described as technological waves and techno-economic paradigms (Knell, 2021). They describe the spread of new clusters of technologies which attract research, development and commercialisation and drastically change the way people live. Starting with Schumpeter's (1939) vision about long waves in capitalist development, this area keeps attracting attention because it analyses fundamental shifts in production, economic and governance processes triggered by innovation (Freeman & Perez, 1988; Pavitt, 1999; Perez, 2010; Nelson et al., 2018). Such technological waves were associated with iron, water and mechanisation (1st wave), steam power, railways and steel (2nd wave), electricity, the internal combustion engine and chemicals (3rd wave), petrochemicals, electronics and aviation (4th wave), digital networks, biotechnology and software development (5th wave) and the current sustainability 6th wave is when we are seeing renewable energy and other environmental technologies taking shape (Hargroves & Smith, 2005; Silva & Di Serio, 2016). What is also interesting is that the 7th technological wave is already at our doorsteps with its focus on healing the planet and people, including fast progress in artificial intelligence and medical technologies, vaccines and devices that improve human health (Marinova et al., 2017). Part of this new 7th wave are the new food-related technologies and smart agriculture.

It is obvious that we can no longer continue with the current ways of food production, particularly with livestock-based products. Alternative proteins are a fast-growing area that covers plant-based options as well as cultured meat, insects and fungi. Agricultural methods also need to change by adopting a circular approach. We explore this in more details in the chapters to follow. All these developments are very exciting, promising and offer high potential to find clever solutions for the current planetary emergency by improving planetary health and facilitating better diets.

Better Calorie Intakes

Currently, there are vast differences in the calorie intakes between countries and across different social groups. The share of people who are hungry without adequate access to food calories has been on the increase since 2015. It reached 8.9% of the global population or 690 million people in 2019, 10 million more than in 2018 and 60 million more than in 2015 (UN, n.d.). Furthermore, 2 billion people did not have regular access to safe, nutritious and sufficient food in 2019 (UN, n.d.).

At the other end of the spectrum are those who are obese and overweight (see Fig. 1.1). Nearly 39% of adults over 18 years of age and 6% of children at pre-school age (below 5 years of age) were overweight in 2019 (Statista, 2021). The share of the US adult population who was overweight was 67% in 2020, with 42.2% of it being obese and 19.3% of young people aged between 2 and 19 being obese (Trust for America's Health, 2021). In many US states, more than 70% of the population was overweight – 72.7% in Mississippi, 72% in West Virginia, 71.8% in Kentucky and 71.4% in Oklahoma (Trust for America's Health, 2021). Similar rates are observed in Australia in 2018 where 67% (75% of men and 60% of women) of the adult population were overweight or obese, including 31% of those being obese, with this share reaching 25% for children and adolescent aged between 5 and 17 (AIHW, 2020). Common consequences from excessive weight are cardiovascular diseases, such as heart disease and stroke, diabetes type 2, now manifested also in children, musculoskeletal disorders, particular osteoarthritis, and some cancers, such as colon, endometrial, breast, ovarian, prostate, liver, gallbladder and kidney

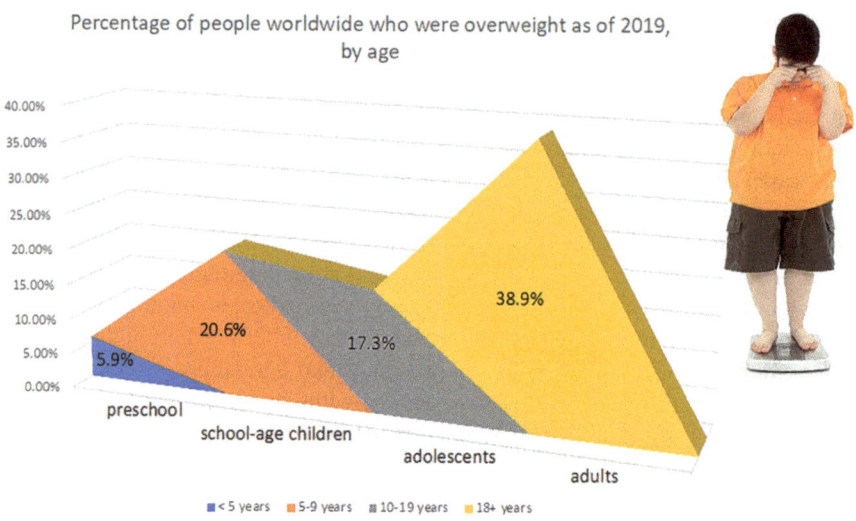

Percentage of people worldwide who were overweight as of 2019, by age

- 5.9%
- 20.6%
- 17.3%
- 38.9%

preschool
school-age children
adolescents
adults

■ < 5 years ■ 5-9 years ■ 10-19 years ■ 18+ years

Fig. 1.1 Worldwide share of people who are overweight, 2019. (Source: Authors, source of data Statista, 2021)

(WHO, 2020). Excessive weight is also an indicator of inappropriate use of food which rather than sustain human existence triggers serious and often life-threatening medical conditions. Australia and USA are also two of the countries where the consumption of animal-based foods, particularly meat, is the highest in the world.

Some populations are exposed to the double burden of poor calorie intake. These are mainly low- and middle-income countries where undernourishment and obesity coexist, particularly within urban settings (WHO, 2020). This can also be the case in wealthier countries, such as USA, when households feel food insecure (FRAC, 2015). Such households are usually in low-income neighbourhoods where there are no farmers' markets or other places where residents can buy cheap locally grown and nutritious vegetables, fruits and whole grains (FRAC, 2015). As data on achieving SDG2 Zero Hunger show, 26% of the global population is affected by moderate or severe food insecurity (UN, n.d.).

Availability of grains for direct human consumption is affected by factors, such as demand from the livestock industry and biofuels, which push farmers to grow for and sell to these markets (GRDC, 2020). Instead of growing and supplying cereals, pulses and seeds as healthy and nutritional food choices, farmers are producing soybeans as animal feed and maize for ethanol production (GRDC, 2020) where their financial benefits are likely to be higher. This essentially increases food insecurity and competition for the use of arable land creating further disparities in the ability to access and make better nutritional choices. Although there may be some decrease in the need for ethanol as a transport fuel and other industrial use due to the COVID-19 pandemic, world demand for corn and soybean destined as animal feed is expected to continue to rise making competition for arable land even more fierce (Lock, 2021). The global consumption of coarse grains is expected to reach

1.5 billion tonnes in 2021–2022 (Lock, 2021) and only 41% are used for direct human consumption while 45% are fed to livestock animals (FAO, 2003). Although some argue that the share of animal feed in the case of global production of cereals is lower at 33% and also that 86% of the matter eaten by animals is not suited for direct human consumption, e.g. grass, leaves, crop residues and fodder (Mottet et al., 2007), the reality is that livestock also competes with wilderness. With 27% of the landmass on the planet taken up by livestock grazing and feed, a further 7% used as cropland for direct human consumption, 10% being glaciers and another 19% barren land, including deserts, salt flats and rocks (Ritchie & Roser, 2019), wildlife and native vegetation are in a losing competition with the human species and its preference for animal-based foods. The scale and rate of biodiversity loss is then easy to explain. To give wilderness a chance we need to transition to better diets with a much higher intake of plant-based foods.

From a planetary health perspective, the better calorie intake applies not only to humans but also to other wildlife species who currently suffer from habitat destruction and land, water and soil pollution induced by the livestock sector. Better food choices in places and communities which have reasonable access to plant-based options would decrease the pressure on the planet's resources. The EAT-Lancet Commission report (Willett et al., 2019) recommends an appropriate calorie intake of around 2500 kilocalories per day with a diversity of plant-based foods, low amounts of animal-source foods, unsaturated rather than saturated fats, and small amounts of refined grains, highly processed foods and added sugars. This Commission as well as all other reputable dietary guidelines recommend limiting the intake of red meat and avoiding processed foods, such as bacon and sausages.

More Sustainable Plant-Based Options

A diet that includes a higher share of plant-based options, or a plant-rich diet, is generally more sustainable. By avoiding animal-based foods, we significantly reduce the demand on land and marine environments, and any associated negative ecological consequences, such as pollution, overfishing and greenhouse gas emissions. Such choices also have an ethical dimension which recognises that animals, including livestock, fish and cephalopods, such as octopus and squid, also have complex nervous systems that can experience pain, fear, stress and distress (Lynch, 2014). Livestock in particular have high levels of intelligence, consciousness and self-consciousness which they manifest, for example, in the ways they look after their young ones. Slaughter, exploitation for milk and eggs are highly traumatic experiences for animals confined in small spaces, separated from their babies and exposed to the cruelty of the meat-producing practices (D'Silva, 2016). Increasing the presence of plant-based choices is also good for human health, a conclusion demonstrated time and again in numerous scientific studies and meta-studies' reviews. We explore further the health and environmental benefits of reduced meat consumption in the chapter on flexitarianism which argues that increased intake of plant-based foods should be mainstreamed in order to transition towards better diets.

Conclusion

The current trends in food production and consumption show that global diets are neither sustainable nor healthy (Fanzo, 2019) which challenges not only the achievement of the Paris Agreement, SDG2 Zero Hunger, SDG14 Life below Water and SDG15 Life on Land, but compromises the entire notion of sustainability. Our food systems are degrading the health of the natural environment, both on land and at sea creating air, soil and water pollution (WRI, 2018; IPCC, 2019; Willett et al., 2019). Humanity is doing a poor job in protecting the biophysical world on which it depends for food and survival (Sachs, 2015).

We do not have other planets where we can expand our environmental footprints and before we all irreversibly fail the human race, its civilisation and future generations, we need to make serious efforts to transition towards better diets. The immediate benefits for individual and public health are much smaller by comparison to the bigger picture of preserving planetary health and giving a chance for the planet to heal and repair itself. We truly believe that the sooner livestock-based foods are put in the attic of history, the greater the probabilities for better, healthier and tastier food opportunities to emerge. They will give and sustain hope in the creativity and values people have. Such a task is very urgent and our book is a modest attempt to present some ideas how this could be achieved.

References

Australian Institute of Health and Welfare (AIHW). (2020). *Overweight and obesity*. https://www.aihw.gov.au/reports/australias-health/overweight-and-obesity

Cedamia. (2021). *Climate emergency declarations*. https://www.cedamia.org/global/

Crippa, M., Solazzo, E., Guizzardi, D., Monforti-Ferrario, F., Tubiello, F. N., & Leip, A. (2021). Food systems are responsible for a third of global anthropogenic GHG emissions. *Nature Food, 2*, 198–209. https://doi.org/10.1038/s43016-021-00225-9

D'Silva, J. (2016). Impact of meat consumption on the health of people, the animals they eat, and the earth's resources. In T. Raphaely & D. Marinova (Eds.), *Impact of meat consumption on health and environmental sustainability* (pp. 201–220). IGI Global.

Department of Economic and Social Affairs (DESA). (2019). *World population prospects 2019: Highlights*. United Nations. https://population.un.org/wpp/Publications/Files/WPP2019_Highlights.pdf

Department of Economic and Social Affairs (DESA). (n.d.). *Goals 15: Protect, restore and promote sustainable use of terrestrial ecosystems, sustainably manage forests, combat desertification, and halt and reverse land degradation and halt biodiversity loss*. https://sdgs.un.org/goals/goal15

Dietary Guidelines Advisory Committee. (2015). *Scientific Report of the 2015 Dietary Guidelines Advisory Committee: Advisory Report to the Secretary of Health and Human Services and the Secretary of Agriculture.* U.S. Department of Agriculture, Agricultural Research Service, Washington, DC. https://health.gov/sites/default/files/2019-09/Scientific-Report-of-the-2015-Dietary-Guidelines-Advisory-Committee.pdf

Edgar, G. J., Langhammer, P. F., Allen, G., Brooks, T. M., Brodie, J., Crosse, W., … Mugo, R. (2008). Key biodiversity areas as globally significant target sites for the conservation of marine biological diversity. *Aquatic Conservation: Marine and Freshwater Ecosystems, 18*(6), 969–983. https://doi.org/10.1002/aqc.902

Fanzo, J. (2019). Healthy and sustainable diets and food systems: The key to achieving Sustainable Development Goal 2? *Food Ethics, 4,* 159–174. https://doi.org/10.1007/s41055-019-00052-6

Food and Agriculture Organisation of the United Nations (FAO). (2003). Basic foodstuffs. In *Agricultural commodities: Profiles and relevant WTO negotiating issues.* http://www.fao.org/3/Y4343E/y4343e02.htm

Food and Agriculture Organisation of the United Nations (FAO). (2012). *Sustainability diets and biodiversity: Directions and solutions for policy, research and action.* http://www.fao.org/3/i3004e/i3004e.pdf

Food and Agriculture Organisation of the United Nations (FAO). (2018). *The State of World Fisheries and Aquaculture 2018 – Meeting the sustainable development goals.* https://www.un-ilibrary.org/content/books/9789210472340#overview

Food and Agriculture Organisation of the United Nations (FAO) & United Nations Environment Programme (UNEP). (2020). *The State of the World's Forests.* http://www.fao.org/3/ca8642en/CA8642EN.pdf

Food Research & Action Centre (FRAC). (2015). *Understanding the connection: Food insecurity and obesity.* https://frac.org/wp-content/uploads/frac_brief_understanding_the_connections.pdf

Freeman, C., & Perez, C. (1988). Structural crisis of adjustment, business cycles and investment behaviour. In G. Dosi, C. Freeman, R. Nelson, G. Silverberg, & L. Soete (Eds.), *Technical change and economic theory* (pp. 38–66). Pinter.

Global Footprint Network. (n.d.). *What is your ecological footprint?* https://www.footprintcalculator.org/

Global Panel on Agriculture and Food Systems for Nutrition (GPAFSN). (2016). *Food systems and diets: Facing the challenges of the 21st century.* https://www.glopan.org/foresight1/

Global Panel on Agriculture and Food Systems for Nutrition (GPAFSN). (2020). *Future food systems: For people, our planet, and prosperity.* https://foresight.glopan.org

Gonzales Fisher, C. M., & Garnett, T. (2016). *Plates, pyramids, planet. Developments in national healthy and sustainable dietary guidelines: A state of play assessment.* Food and Agriculture Organisation of the United Nations and Oxford University. http://www.fao.org/3/I5640E/i5640e.pdf

Grain Research & Development Corporation (GRDC). (2020). *Industry at a glance.* https://rdeplan.grdc.com.au/industry-at-a-glance

Hallström, E., Carlsson-Kanyama, A., & Börjesson, P. (2015). Environmental impact of dietary change: A systematic review. *Journal of Cleaner Production, 91,* 1–11. https://doi.org/10.1016/j.jclepro.2014.12.008

Hargroves, K., & Smith, M. (2005). Natural advantage of nations. In K. Hargroves & M. Smith (Eds.), *The natural advantage of nations: Business opportunities, innovation, and governance in the 21st century* (pp. 7–33). Earthscan.

Harvard University T. H. Chan School of Public Health. (2021). *Sustainability.* https://www.hsph.harvard.edu/nutritionsource/sustainability/

Hauver, E. (2019). *The inconvenient truths behind the 'Planetary Health' diet.* https://www.greenbiz.com/article/inconvenient-truths-behind-planetary-health-diet

Intergovernmental Panel on Biodiversity and Ecosystem Services (IPBES). (2019). *Summary for Policymakers of the IPBES Global Assessment Report on Biodiversity and Ecosystem Services.* https://www.ipbes.net/global-assessment

Intergovernmental Panel on Climate Change (IPCC). (2019). *Climate change and Land*. Special report on climate change, desertification, land degradation, sustainable land management, food security, and greenhouse gas fluxes in terrestrial ecosystems. https://www.ipcc.ch/srccl/

International Union for Conservation of Nature (IUCN). (2021). *Nature's future, our future – The world speaks*. https://www.iucn.org/news/protected-areas/202102/natures-future-our-future-world-speaks

Knell, M. (2021). The digital revolution and digitalized network society. *Review of Evolutionary Political Economy, 2*, 9–25. https://doi.org/10.1007/s43253-021-00037-4

Lebreton, L., Slat, B., Ferrari, F., Sainte-Rose, B., Aitken, J., Marthouse, R., ... Reisser, J. (2018). Evidence that the Great Pacific Garbage Patch is rapidly accumulating plastic. *Scientific Reports, 8*, 4666. https://doi.org/10.1038/s41598-018-22939-w

Lock, P. (2021). *Coarse grains: March quarter 2021*. Department of Agriculture, Water and the Environment. Australian Government. https://www.agriculture.gov.au/abares/research-topics/agricultural-outlook/coarse-grains

Lynch, K. (2014). *When is an animal not an 'animal'? Research ethics draws the line*. The Conversation. https://theconversation.com/when-is-an-animal-not-an-animal-research-ethics-draws-the-line-21756

Marinova, D., Hong, J., Todorov, V., & Guo, X. (2017). Understanding innovation for sustainability. In J. Hartz-Karp & D. Marinova (Eds.), *Methods for sustainability research* (pp. 217–230). Edward Elgar.

Marris, E. (2019). *To keep the planet flourishing, 30% of Earth needs protection by 2030*. National Geographic. https://www.nationalgeographic.com/environment/article/conservation-groups-call-for-protecting-30-percent-earth-2030

Milo-Dale, L. (2020). *Europe is failing to implement the United Nations' (UN) 2030 Agenda*. https://www.wwf.eu/?uNewsID=360550

Mottet, M., de Haan, C., Falcucci, A., Tempio, G., Opio, C., & Gerber, P. (2007). Livestock: On our plates or eating at our table? A new analysis of the feed/food debate. *Global Food Security, 14*, 1–8. https://doi.org/10.1016/j.gfs.2017.01.001

Myers, S., & Frumkin, H. (Eds.). (2020). *Planetary health: Protecting nature to protect ourselves*. Island Press.

Nelson, R., Dosi, G., Helfat, C., Pyka, A., Saviotti, P. P., Lee, K., & Winter, S. G. (2018). *Modern evolutionary economics: An overview*. Cambridge University Press.

Pavitt, K. (1999). *Technology, management and systems of innovation*. Edward Elgar.

Perez, C. (2010). Technological revolutions and techno-economic paradigms. *Cambridge Journal of Economics, 34*(1), 185–202. https://doi.org/10.1093/cje/bep051

Peters, C. J., Picardy, J., Darrouzet-Nardi, A. F., Wilkins, J. L., Griffin, T. S., & Fick, G. W. (2016). Carrying capacity of U.S. agricultural land: Ten diet scenarios. *Elementa: Science of the Anthropocene, 4*, 000116. https://doi.org/10.12952/journal.elementa.000116

Poore, J., & Nemecek, T. (2018). Reducing food's environmental impacts through producers and consumers. *Science, 360*(6392), 987–992. https://doi.org/10.1126/science.aaq0216

Ritchie, H., & Roser, M. (2019). *Land use*. Our World in Data. https://ourworldindata.org/land-use

Sachs, J. D. (2015). *The age of sustainable development*. Columbia University Press.

Schumpeter, J. A. (1939). *Business cycles: A theoretical, historical, and statistical analysis of the capitalist process*. McGraw-Hill.

Shapiro, P. (2016). Feasting from the federal trough: How the meat, egg and dairy industries gorge on taxpayer dollars while fighting modest rules. In T. Raphaely & D. Marinova (Eds.), *Impact of meat consumption on health and environmental sustainability* (pp. 244–254). IGI Global.

Silva, G., & Di Serio, L. C. (2016). The sixth wave of innovation: Are we ready? *RAI Revista de Administração e Inovação, 13*(2), 128–134.

SolAbility. (2020). *The Global Sustainable Competitiveness Index 2020*. https://solability.com/the-global-sustainable-competitiveness-index/the-index

Statista. (2021). *Percentage of people worldwide who were overweight as of 2019, by age*. https://www.statista.com/statistics/1065605/prevalence-overweight-people-worldwide-by-age/

Steffen, W., Richardson, K., Rockström, J., Cornell, S. E., Fetzer, I., Bennett, E. M., … Sörlin, S. (2015). Planetary boundaries: Guiding human development on a changing planet. *Science, 347*(6223), 736. https://doi.org/10.1126/science.1259855

Tirado, R., Thompson, K. F., Miller, K. A., & Johnston, P. (2018). *Less is more: Reducing meat and dairy for a healthier life and planet.* Greenpeace Research Laboratories Technical Report (Review) 03–2018. https://storage.googleapis.com/planet4-international-stateless/2018/03/6942c0e6-longer-scientific-background.pdf

Tollefson, J. (2021). COVID curbed carbon emissions in 2020 – But not by much. *Nature, 589*(7842), 343.

Trust for America's Health. (2021). *The state of obesity 2020: Better policies for a healthier America.* https://www.tfah.org/report-details/state-of-obesity-2020/

United Nations (UN). (n.d.). *Sustainable Development Goals. Goal 2: Zero Hunger.* https://www.un.org/sustainabledevelopment/hunger/

United Nations Environment Programme World Conservation Monitoring Centre (UNEP-WCMC), International Union for Conservation of Nature (IUCN) & National Geographic Society (NGS). (2021). *Protected Planet Live Report 2021.* UNEP-WCMC, IUCN and NGS: Cambridge, UK. https://livereport.protectedplanet.net

United Nations Framework Convention on Climate Change (UNFCCC). (2021). *Planetary health.* https://unfccc.int/climate-action/momentum-for-change/planetary-health

Willett, W., Rockström, J., Loken, B., Springmann, M., Lang, T., Vermeulen, S., … Murray, C. J. L. (2019). Food in the Anthropocene: The EAT–Lancet Commission on healthy diets from sustainable food systems. *The Lancet, 393*(10170), 447–492. https://doi.org/10.1016/S0140-6736(18)31788-4

World Commission on Environment and Development. (1987). *Our common future.* https://sustainabledevelopment.un.org/content/documents/5987our-common-future.pdf

World Health Organisation (WHO). (2020). *Obesity and overweight.* https://www.who.int/news-room/fact-sheets/detail/obesity-and-overweight

World Resources Institute (WRI). (2018). *Creating a sustainable food future: A menu of solutions to feed nearly 10 billion people by 2050.* https://files.wri.org/s3fs-public/creating-sustainable-food-future_2.pdf

World Wide Fund for Nature (WWF) Europe (EU). (2019). *EU Overshoot Day: Living beyond nature's limits.* http://awsassets.panda.org/downloads/wwf_eu_overshoot_day___living_beyond_nature_s_limits_web.pdf

Worldometer. (2021). *World population by year.* https://www.worldometers.info/world-population/world-population-by-year/

Chapter 2
Climate and Food

Abstract Given the complexity of the connections between food and climate change, this chapter describes the links between greenhouse gas emissions, extreme weather events and food production. It also specifically analyses the impact of industrial fishing on the ability of oceans to store carbon. Vulnerability of food production to climate change and extreme weather events, and the vulnerability of people to changes in food production are further examined. The overall message in the current planetary emergency is that a transition to plant-based options is a good strategy for responding to the climate change crisis.

In November 2019, we were at a food conference in Taichung, Taiwan (see Fig. 2.1). It was an amazing event and splendid time of face-to-face meetings and productive in-person discussions about different aspects of food. The COVID-19 pandemic was yet to emerge and change all corners of the world. There was something else however that came as a shock to us as we were listening to the presentations in the Taichung conference hall – the bushfires in Australia. We were glued to the news in disbelief that Australia was burning so early – it was only spring. The scale of the fires was enormous, devastating and unprecedented. Later on, we would find out that 186,000 square kilometres (46 million acres or 72,000 square miles) have burnt (CDP, 2020). Such a fire would have erased to the ground our native Bulgaria – a country with a territory of 110,000 square kilometres. So many other countries are smaller than the area decimated by the 2019–2020 Australian bushfires – Cambodia, Uruguay, Tunisia, Bangladesh, Greece and the list goes on…

The bushfires released 350 million tonnes of CO_2 into the atmosphere (Sanderson & Fisher, 2020) which is higher than the individual annual emissions of high-emitting countries, such as Poland, France, Italy and Kazakhstan (UCS, 2020). Other estimates put that figure at between 400 and 700 million tonnes of CO_2 (Bishop, 2020) which makes it comparable to the annual emissions of Australia or UK (UCS, 2020). In addition to the thousands of houses, buildings and other infrastructure destroyed in the bushfires and the loss of lives of several firefighters, there were more than 3 billion native vertebrate animals who perished in the inferno

D. Marinova, D. Bogueva, *Food in a Planetary Emergency*, https://doi.org/10.1007/978-981-16-7707-6_2

19

Fig. 2.1 Rainbow Village, Taichung, Taiwan, 2019 – simple actions that changed the destiny of a place. (Source: Authors, Prof. Dora Marinova – left, and Dr Diana Bogueva – right)

(WWF Aus, 2020), including 143 million mammals, 180 million birds, 51 million frogs and 2.46 billion reptiles (the latter living in much higher densities in the fire-affected areas). The Australian World Wide Fund for Nature describes the 2019–2020 Australian bushfires as "one of the worst wildlife disasters in modern history" with no other calamity anywhere in the world in living memory killing or displacing that many animals (WWF Aus, 2020, p. 1).

Later on, Australia's Commonwealth Scientific and Industrial Research Organisation (CSIRO, 2021) would also explain that although climate change does not cause fires directly, it significantly increases the occurrence of extreme fire weather and lengthens the fire season. High temperature levels and low humidity are two of the main factors contributing to weather conditions conducive to wildfires. Since weather records began in Australia (in 1900 for rainfall and 1910 for temperature), 2019 – when the bushfires commenced, held simultaneously two records by being the continent's driest and warmest year with an annual mean temperature

1.52 °C above the average (CSIRO, 2021). The year that followed – 2020, was the fourth warmest year on record (BoM, 2021). In fact, all years since 2013 have been amongst the warmest on record for the Australian continent (BoM, 2021).

Wildfires are one of the most complex whether-related extreme events to understand and analyse (Sanderson & Fisher, 2020). Notwithstanding this, a modelling study by van Oldenborgh et al. (2021) found a clear connection between anthropogenic climate change and fire weather in Australia with extreme heats becoming at least twice more likely due to the long-term warming trend.

There was something else that was also disturbing at the 2019 food conference in Taiwan – nobody except us spoke about the link between food and climate change. One of the most serious anthropogenic factors contributing to the current climate and environmental emergency was largely left ignored. Despite overwhelming evidence that food production is a direct contributor to climate change, the silos between these two areas remain in academia as well as amongst the general public. This chapter tries to address the lack of general knowledge and acceptance that the food choices we make have a direct and large environmental footprint in terms of greenhouse gas emissions and are seriously contributing to climate change.

On a 20-year assessment period, the global warming potential of agriculture represents 22% of all sector contributions (IPCC, 2014b) and according to these 2014 estimates, is higher than that of electricity and hear production (17%), other energy, such as petroleum refining, coke ovens, briquettes, blast furnaces for pig iron production (17%), other industry (20%), transportation (10%), buildings (6%) and forestry and other land use (8%). There are many reasons why we do not hear about food's defining contribution to climate change. One of them is purely technical. Most calculations of carbon dioxide equivalency are conducted over 100 years. However, once released methane is a greenhouse gas with a relatively short-term presence in the Earth's atmosphere, namely around 25 years, while carbon dioxide remains unchanged for hundreds of years (so does nitrous oxide). By using a global warming potential over 100 years instead of 20 years, the climate-forcing effect of methane is significantly diminished and its CO_2 equivalency drops to 28 compared to 84 (IPCC, 2014b). From an industry perspective methane is largely associated with agriculture and in particular, livestock production. Hence, in any assessments that use a 100-year time horizon, the weight assigned to food production is significantly reduced and this has commonly been the case (Wedderburn-Bisshop & Rickards, 2018). Such a technical approach has had many policy implications which have resulted in an emphasis on prioritising addressing carbon dioxide, and related industries, over methane and food production.

Given the complexity of the connections between food and climate change, this chapter tries to shed light on the links between greenhouse gas emissions, extreme weather events and food production. We also specifically analyse the impact of industrial fishing on the ability of oceans to store carbon. Vulnerability of food production to climate change and extreme weather events, and the vulnerability of people to changes in food production are then discussed. The overall message in the current planetary emergency is that a transition to plant-based options is a good strategy for responding to the climate change crisis.

Food and Greenhouse Gas Emissions

Agriculture in all its shapes and forms dominates food production. As of 2013, "83% of the 697 kg of food consumed per person per year, 93% of the 2884 kcal per day, and 80% of the 81 g of protein eaten per day" were coming from production on the land (IPCC, 2019, p. 445). Between 1961 and 2016, the greenhouse gas emissions associated with food production have grown from 3.1 to 5.8 Giga tonnes CO_2-equivalent per year (IPCC, 2019) with the rise mainly driven by the livestock sector – through enteric fermentation and manure left on pasture responsible for the release of methane, increased use of chemical fertilisers – associated with nitrous oxide, and to a certain extent rice cultivation – rice sequesters carbon but the flooded rice paddocks emit methane because of decomposing organic matter. In Australia, per each tonne of nitrogen fertiliser, 25 kg are lost in the atmosphere, where it forms nitrous oxide combining with oxygen (Grace & Barton, 2014).

When estimated over a 100-year equivalency, recent estimates show that between 21% and 37% of the total global greenhouse gas emissions are attributed to the current food-related activities, including agriculture, land use change, storage, transport, packaging, processing, retailing and consumption (IPCC, 2019). This contribution would be even higher if the methane associated with livestock and fertilisers is estimated over a 20-year period, making agriculture the sector which impacts climate change by far the most. The IPCC's 6th Assessment Report also states that agriculture dominated by livestock is a driver of the faster growth of methane emissions during 2014–2019.

The need to intensify crop production to satisfy the demand for food and animal feed has resulted in an 800% increase in the application of nitrogen fertilisers between 1961 and 2016 (IPCC, 2019) which contribute towards the release in the atmosphere of nitrous oxide, a greenhouse gas 265 times more powerful than CO_2. There is also increasing empirical evidence that some fertiliser plants, which commonly use natural gas as a feedstock and a fuel for the production of ammonia, emit 100 times more methane than officially reported (Zhou et al., 2019). Without the need to continuously feed increasing numbers of livestock animals, the pressure to intensity the production of feed crops using fertilisers, and to convert native areas of vegetation into grazing and agricultural land would also reduce.

Compared to plant-based options, ruminant meats have 20–100 times higher greenhouse gas emission impacts (Clark & Tilman, 2017). They are also the most inefficient way of providing proteins compared to other meats, milk, eggs and all crops (IPCC, 2014b, 2019). Beef in particular is the most detrimental environmental option according to a meta-analysis of 570 studies from 119 countries examining the life-cycle environmental footprints of 40 different food items which provide 90% of the global protein and calorie intake (Poore & Nemecek, 2018). Figure 2.2 shows the greenhouse gas emissions across the supply chain associated with one kilogram of different foods. The majority of the emissions are generated in the farm and because of land use change (Ritchie & Roser, 2021). In the climate emergency,

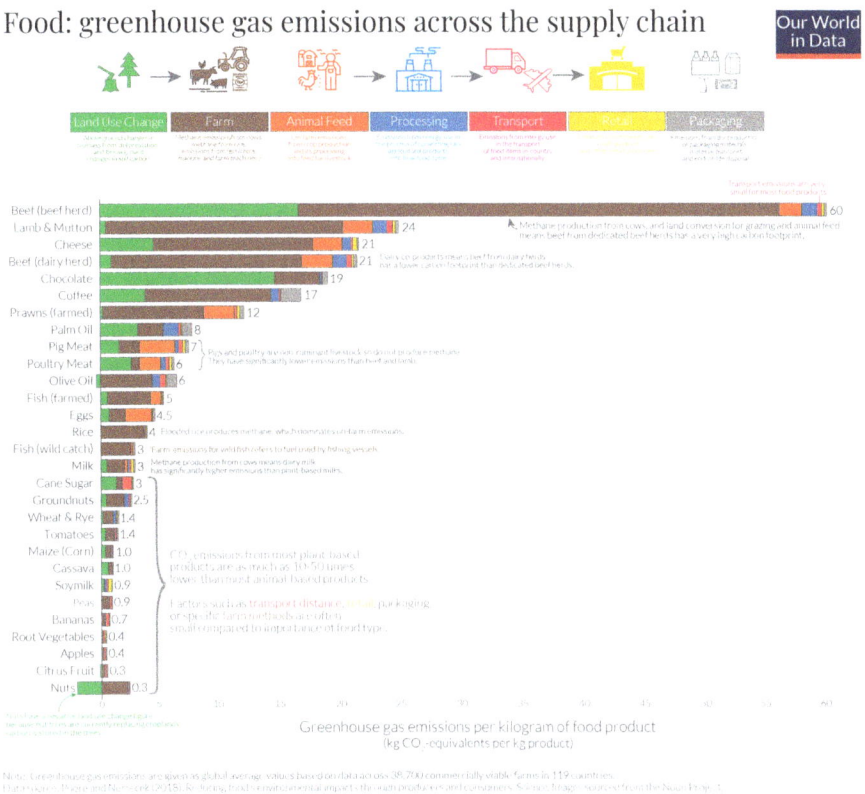

Fig. 2.2 Greenhouse gas emissions across the supply chain associated with different food. (Source: Ritchie & Roser, 2021)

reduction in the intake of ruminant meats seems the first most obvious step to take to cut greenhouse emissions.

Numerous studies have compared the greenhouse gas emissions of diets which exclude meat (Marinova & Raphaely, 2018). Although data varies according to geographic regions and individual countries, the evidence is clear that replacing meat with plant-based options can cut the greenhouse gas emissions significantly – from 29% in the case of North America (Soret et al., 2014) to 34% for vegetarian and 50% for vegan options in the UK (Scarborough et al., 2014) to 45% less if a meal with mutton is replaced with a vegetarian option in India (Pathak et al., 2010). The EAT-Lancet Commission (Willett et al., 2019) planetary health diet recommends limiting significantly the intake of animal-based foods in order to achieve the Sustainable Development Goals (SDGs) and the Paris Agreement of limiting temperature rises to 2 °C, preferably 1.5 °C. The recommended daily quantities for a 2500 kcal per day diet are 14 g (between 0 and 28 g) of red meat (beef, lamb and pork), 29 g (between 0 and 58 g) of poultry, 13 g (between 0 and 25 g) of eggs and 28 g (between 0 and 100 g) for fish (Willett et al., 2019). Vegetarian and vegan diets are now fully supported by doctors and environmentalists alike as legumes and nuts are excellent alternative protein sources that are much more beneficial to combat climate change and respond to the planetary emergency.

People who consume meat and other animal-based products, particularly in quantities larger than the ones advised by the EAT-Lancet Commission, contribute disproportionately more to the current level of global greenhouse gas emissions. These people also tend to belong to countries and sections of society where malnourishment and malnutrition prevail because of deficiencies, excesses and imbalances in the intake of nutrients (IPCC, 2019). The estimated number of these people is 2 billion adults who are also overweight and obese (IPCC, 2019). At the other end of the spectrum are the 800 million people who are undernourished and their food consumption is insufficient to maintain a normal healthy life. Equity in food consumption implies looking after the interests of all people on this planet – current and future generations. A diet rich in plant-based options is much better suited to deliver equity in consumption. It can help avoid the devastating impacts climate change could have on the socially weak people within society and the global economy.

Fish and Carbon Storage

Marine fisheries (or captured fisheries) are commonly excluded from global assessments of greenhouse gases associated with food (Parker et al., 2018). There have however been a few estimates as to how much fishing contributes to climate change through fuel consumption and emissions of the global fleet. For example, Parker et al. (2018) estimate that in 2011 fishing contributed 4% of all greenhouse gas emissions associated with food production amounting to 179 million tonnes of CO_2-equivalent. They explain that emissions from fishing are on the rise in absolute terms – a 28% increase between 1990 and 2011, and also, without simultaneous increase in productivity, that is emissions per tonne of fish caught have grown by 21% during the same period (Parker et al., 2018). The main reasons are expansion in fishing activities and increased harvesting from fuel-intensive species, such as crustaceans. Australia's fishing fleet is particularly fuel-intensive as it targets crustaceans, such as rock lobsters and prawns. Although Australia accounts for only 6% of global seafood production, it generates over 22% of the associated greenhouse gas emissions (Farmery et al., 2018).

Compared to other animal proteins, products derived from marine fisheries were believed to be associated with lower carbon emissions, similar to those in poultry production (Parker et al., 2018), but still higher than most plant-based foods. A 2020 study adds an extra dimension to the consumption of fish, particularly large species such as tunas, sharks, mackerels and billfishes (marlins, swordfish, spearfish and sailfish). While most terrestrial organisms release carbon in the atmosphere after they die, "carcasses of large marine fish sink and sequester carbon in the deep ocean" for thousands of years (Mariani et al., 2020, p. 1). When such fish are being caught, their processing and consumption contribute towards additional atmospheric emissions. When the disruption of this natural carbon sink is taken into consideration, the emissions produced through fishing are 25% higher than the estimates based only on fuel consumption (Mariani et al., 2020). For example, the 2011 estimate of 179 becomes 224 million tonnes of CO_2-equivalent, a difference that is far from being negligible. Hence, industrial fishing interrupts largely the ocean's blue carbon pump. As marine fisheries have depleted most fish stocks compared to preindustrial levels (Costello et al., 2016), government subsidies enable "fishing fleets to travel vast distances and burn large amounts of fossil fuel to reach remote fishing grounds in the high seas", supporting an industry that continues to overexploit the oceans and is responsible for large quantities of greenhouse gas emissions (Mariani et al., 2020, p. 1).

Industrial fishing also catches smaller pelagic species which are used as fishmeal in aquaculture and feed for livestock. Although these species have a smaller carbon footprint compared to those directly for human consumption, they contribute towards the overall high carbon emissions associated with animal-based proteins. Furthermore, the global emissions from aquaculture are higher than those of commercial fishing and are estimated to represent 250 million tonnes of CO_2-equivalent in 2017 (MacLeod et al., 2020). Aquaculture's emission intensity is also higher than that for marine fisheries, is comparable to pork (MacLeod et al., 2020) and is higher than most plant-based food options.

Commercial trawling is particularly detrimental to the marine environment as it affects the ecological balance of species and generates a lot of plastic pollution. If marine fisheries are better managed with scientifically-based harvesting policies and reforms aligning profits with conservation, it would take 10 years for them to recover from the current status of overfishing (Costello et al., 2016). There is hence the expectation that if wild-caught landings are restricted to protect the health of the marine environments, aquaculture would need to significantly grow to satisfy higher demands for fish (Tilman & Clark, 2014). Globally, aquaculture already supplies more fish than marine fisheries (IPCC, 2019). A further transition to more aquaculture would come at a higher ecological cost for the planet's climate.

Extreme Weather Events and Food

In the previous sections we discussed how food production is contributing to greenhouse emissions which are the main cause for anthropogenic climate change. Science of climate change has advanced so much that nowadays it can detect global warming on the basis of any single day observations of temperature and moisture based on a global spatial pattern (Sippel et al., 2020). This also allows to attribute the contribution of externally forced changes on the frequency and magnitude of extreme weather events. There is no hesitation whatsoever left that climate change is externally driven by human activities, with food production being a major component of this.

This climate change is manifested through extreme weather events. The 2019–2020 Australian fires were one such event. Since 1980, the number of floods has quadrupled; extreme temperatures, droughts and wildfires, have more than doubled, and storms have doubles (European Academies Science Advisory Council, 2018). The United Nations Office for Disaster Risk Reduction (UNODRR, 2020) describes a "staggering rise" in the climate-related disasters around the world in the first 20 years of the twenty-first century. During 2000–2019, there was a total of 7348 major disasters, affecting 4 billion people, claiming 1.23 million human lives and resulting in US$2.97 trillion in economic losses (UNODRR, 2020). By comparison, the previous 20 years, 1980–1999, recorded 4212 major disasters, affecting 3 billion people, claiming 1.19 million human lives and causing US$1.63 trillion economic losses (UNODRR, 2020).

The COVID-19 pandemic, which commenced in 2020, changed our perception about disasters and how prepared we are to deal with them. By the end of 2020, the pandemic claimed 2 million lives and caused trillions of dollars in economic losses (UNODRR, 2021). The climate-related disasters in 2020 pale by comparison with the COVID-19 devastation. However, they were significant. The year 2020 equalled the 2016 record in being the world's hottest year on record; there were 389 major climate-related disaster events with 98.4 million people affected, 15,080 human deaths and economic losses of US$ 171.3 billion (UNODRR, 2021). Although the number of people affected and human lives lost were lower than the average for the previous two decades, this was due mainly to the fact that in 2020 there was not a major catastrophic event claiming millions of lives. However, in 2020 there were overall more disaster events with 26% more storms and 23% more floods than the annual average. The size of economic losses was also higher in 2020 compared to the previous average (UNODRR, 2021). While the world was pre-occupied with the pandemic, climate change continued to be a major factor affecting live on Earth.

Since the 2000s, climate science has been exploring the human fingerprint or attribution to extreme weather events, such as droughts, heatwaves, storms, floods and wildfires (Carbon Brief, 2021). There is mounting evidence that human activity, including food production, is increasing the risk for extreme weather and particularly, heat and drought, in both frequency and magnitude. Out of the 405 extreme weather events analysed by Carbon Brief (2021), 70% were made more likely or more severe by human-caused climate change; in the case of heat the percentage of attribution was 92% and for drought – 65%. As a warmer atmosphere contains more moisture, it generates heavy rainfalls and increases the magnitude of storms. With each degree of global warming, it is projected that the most extreme precipitation events are likely to double in intensity and frequency (Myhre et al., 2019). What does this mean for food production?

Extreme weather events and climate-related disasters are challenging food security in any part of the world. All food production is being impacted. Drought is crippling the growth of crops; floods are damaging yields and seeds; storms are devastating harvests; extreme weather events are destroying agricultural infrastructure and assets; outbreaks of plant pests and diseases are being witnessed with changing climatic conditions (FAO, 2021). All these weather-related events are contributing to food losses at the farm level and triggering price spikes. They also result in competition for crops within the global agricultural sector, for example between grains for feed and for direct human consumption, as well as with the bioenergy sector (Tadesse et al., 2014). The poor and socially weak people are being disproportionately affected by price spikes, particularly as a large fraction of their household budgets is being spent on food (Hossain & Green, 2011).

Not only extreme weather events, persistent climate change is going to impact yields of starchy dietary staples as well as non-staple vegetables and legumes (Scheelbeek et al., 2018). Grain production is likely to experience shocks due to climate-related events and this impacts crops, such as maize, which are used both as food and feed (Tigchelaar et al., 2018). A 2017 assessment of the four major crops that provide two-thirds of the human calorific intake, shows that for every degree Celsius increase in temperature, there will be a reduction in the global yields of wheat by 6.0%, rice by 3.2%, maize by 7.4%, and soybean by 3.1% (Zhao et al., 2017). The production of sorghum, which is a crop with the ability to withstand harsh environmental conditions, is likely to decrease even further – around 10% per each degree of temperature warming (Tack et al., 2017).

As oceans heat up due to global warming, many fish and shellfish migrate to new territories but their bodies are also affected with many unknown implications for dominant and subordinate species (Hollowed et al., 2013). The timing of fish spawning is affected towards earlier months and warm waters favour smaller individuals which is likely to reduce the expected levels of catch (Simpson et al., 2013). A 2021 review of studies on the impacts of climate change on fish populations shows that only 1% of all recorded freshwater and marine fish species (309 of approximately 35,000) have been studied to examine their response (Huang et al., 2021). The overall conclusion is that climate change through temperature impacts has negative effects on fish growth reflected in their physiology and health. Higher temperatures increase the presence of pathogens and diseases and can also speed up fish metabolism resulting in a faster uptake of toxins which can affect human health (Cho, 2018). In addition, ocean acidification resulting from climate change poses higher risks to shellfish fisheries and aquaculture (Weatherdon et al., 2016).

Ironically, the sector which contributes the most to global warming, namely livestock production, is the one most affected by extreme weather events. According to Rojas-Downing et al. (2017, p. 145), "climate change is a threat to livestock production because of the impact on quality of feed crop and forage, water availability, animal and milk production, livestock diseases, animal reproduction, and biodiversity". With climate change, livestock animals are exposed to heat stress which leads to altered metabolism and increased mortality; diseases related to increased pathogens, parasites and vectors, decreased resistance of livestock, new viruses and other diseases as well as shortages in drinking and servicing water (FAO, 2016). There are also decreases in the availability of forages and feed crops which are in turn affected by climate change (FAO, 2016). The labour force and capital associated with the livestock industry are similarly impacted with many people experiencing mental health problems, conflicts over resources and ultimately looking to migrate either to the cities or other areas with a cooler climate (FAO, 2016). Many livestock animals simply die stranded in the floods, burn to death in the wildfires or are euthanised by their owners to save them from suffering. In the 2019–2020 Australian catastrophic wildfires, 125,000 livestock perished (Wahlquist, 2020).

Rivera-Ferre et al. (2016) argue that we should distinguish between different livestock systems, such as grazing, mixed crop-livestock and industrial systems as they are differently being impacted (directly and indirectly) by mean climate changes and extreme events, and their adaptation capacities differ. Adaptation through genetics, nutrition and management is put forward by many researchers in the area of livestock (Bernabucci, 2019). This is certainly the case and there is some progress in trying to decrease the emissions' footprint of livestock through land management (e.g. agroforestry) or feeding additives (such as seaweed to reduce the flatulence in cattle) but the only true alternative is to reduce demand for the consumption of meat and other animal-based products, including dairy, seafood and eggs. Without tackling the most greenhouse gas-intensive foods, such measures may not be sufficient and may also raise other environmental and ethical concerns (Garnett, 2011).

By shifting to more plant-based food options in areas of the world where this is possible, we not only cut direct emissions from the livestock animals but also eliminate the need to grow extra crops as feed. Reduction of agricultural losses and food waste (discussed later in the book) is important, and it should also be acknowledged that wasted animal-based foods have a much higher ecological footprint and contribution to greenhouse gas emissions. Changing diets towards a smaller presence of animal-sourced foods will put less strain on the planet's ecoservices and the global agricultural resources. Shortening the food chain from "plants–to livestock–to humans" to "plants–to humans" (Schmidinger et al., 2018) is a much more efficient logical way to achieving human nutrition and providing food security. It is also a way to respond to the climate emergency.

The Western Australian government already warns farmers and livestock managers that climate change will have increasingly negative effects on their businesses for which they have to prepare (DPIRG, 2021). Possible strategies stated are new technology, operations and management to provide shelter, shade and cooling as well as genetically breeding of species that are more adapted to warmer conditions; the government also acknowledges that a complete transformation in the feed base is likely to be required with the advent of global warming and changes in precipitation patterns. The strategy that is not listed but is most needed, is the one that encourages livestock farmers to look for other opportunities that are less damaging to the planet's climate.

Climate Change and Vulnerability in Food Production

The relationship between food and climate change is bidirectional, it works in both ways. Food production and consumption contribute to changes in the climate and the severity of these changes affects food production. All assessments of the Intergovernmental Panel on Climate Change (IPCC) show that that agriculture is one of the economic sectors most vulnerable to climate change (e.g. IPCC, 2014a, 2021). Crops and livestock are exposed to a wide range of climate-related natural hazards, including drought, flooding, wildfire, sea level rise and soil erosion. The loss of climatic stability is likely to affect farming practices making food production even more vulnerable to climate change across the world, with developing countries having less resources and lower adaptive capacity to respond to such emergencies.

Climate change negatively impacts all aspects of food security, including availability, access, utilisation and stability (IPCC, 2019). Although at relatively low temperature increases, a higher concentration of carbon dioxide in the atmosphere may improve the growth and yields of some crops, it decreases their nutritional value because of lower protein content. The spread of pests and diseases will also change impacting negatively food production (IPCC, 2019). Out of all agricultural activities, the livestock industry and pastoralism in particular are expected to be most affected by climate change. The vulnerability of pastoralism is projected to be very high compared to, for example, medium for fruit and vegetable production (IPCC, 2019). In IPCC's 6th Assessment Report, all regions across the globe are projected to experience increases in hot and decreases in cold climatic impact-drivers influencing agriculture's vulnerability (IPCC, 2021).

This agricultural vulnerability has socio-economic and health, including psychological, consequences. Post-traumatic stress, adjustment disorder and depression are linked to increased frequencies of disasters with climate change (Padhi et al., 2015). Farmers and entire populations are likely to temporarily move or migrate to protect their livelihoods.

The Climate Institute (2011) of Australia describes inaction on climate change as being responsible for the climate of suffering that is already upon us and triggering food vulnerability. New terms are being introduced to convey the seriousness of the impacts climatic events have on people; they include ecoanxiety, ecoguilt, ecological grief, solastalgia and biospheric concern (Cianconi et al., 2020). Professional medical organisations, such as the International Society of Doctors for the Environment (2016), its branch Doctors for the Environment Australia (2020) and the American Psychiatric Association (APA, 2021), recognise the threats climate change poses to human health and the fact that climate is also threatened by human activities. The task before humanity and each country is twofold: "we must manage the unavoidable changes already in the pipeline and, at the same time, avoid the unmanageable human tragedy of climate change unchecked" (The Climate Institute, 2011, p. 5). Acting on shifting our dietary preferences towards more plant-based options is long overdue but a transition now would reduce humanity's vulnerability to climate change and deal with many aspects of our biospheric concerns.

Controlling greenhouse gas emissions and transitioning to carbon neutral economic activities are essential to maintain a stable climate on this planet. As not all industries can be carbon neutral, the concept of Net Zero emerged as a task describing that an overall zero greenhouse gas emissions balance from all human activities is needed to respond to the climate change challenge. Technological options to deal with emissions in agriculture are limited and the task is quite complex, particularly because of its livestock component (Energy & Climate Intelligence Unit, 2021). The way animal-based goods are currently produced generates excessive amounts of greenhouse gas emissions but also requires massive land use change which clears natural carbon sinks, such as forests, peatlands, steppes and savannahs. A reduction in the emissions-heavy foods and activities would ease the achievement of Net Zero pledges by countries, including Finland (by 2035), Iceland, Austria (2040), Sweden (2045), Chile, Costa Rica, Denmark, France, Germany, UK, New Zealand, Norway, Portugal, Spain, Switzerland, USA (2050) and China (2060). Some countries, such as Australia, are aspiring to make their livestock industry carbon-neutral; however, this is a highly challenging task with a lot of unknowns. Most importantly, as we discussed in the first and later chapters of this book, reducing the intake of animal-based foods has numerous human and planetary health co-benefits. This is a realistic strategy with less unknows that depends on people's will.

Conclusion

Our conference in Taiwan was about food but the basic conditions that allow us to grow, catch and produce food were not centre stage. At the same time, the effects of climate change were devastating Australia through the bushfires and were a sign of what is to come. It seems that humanity continues to be wilfully destructive by not making or ignoring the links between the food we produce and eat and its impacts on climate change. The first key finding of the IPCC's 6th Assessment Report is that human activities have caused global warming (IPCC, 2021); its second finding is that agriculture, forestry and land-use changes represented (on the stretched 100-year scale) 23% of the total net anthropogenic greenhouse gas (GHG) emissions during 2007–2016 or 12 Gt of CO_2 equivalent per year, accounting for 13% of CO_2, 44% of methane and 82% of nitrous oxide emissions from all human activities (IPCC, 2021).

However, we still have an option and a choice to avoid doom-and-gloom future scenario and flatten the climate disaster curve. Changing human preferences towards higher intake of plant-based foods offers a big opportunity in the tightly linked diet–climate change dilemma. Some animal-based options, such as fish, poultry and pork may be better for the climate than ruminant meats, such as beef/veal, lamb/ mutton and goat; however, plant-based foods are superior as they generate less emissions during the production process, can also store carbon and produce oxygen through the process of photosynthesis. Nuts are particularly good for storing carbon. The Intergovernmental Panel on Climate Change (2019, p. 519) describes succinctly how changing diets can help climate change:

> changing diets towards a lower share of animal-sourced food, once implemented at scale, reduces the need to raise livestock and changes crop production from animal feed to human food. This reduces the need for agricultural land compared to present and thus generates changes in the current food system. From field to consumer this would reduce overall GHG emissions. Changes in consumer behaviour beyond dietary changes, such as reduction of food waste, can also have, at scale, effects on overall GHG emissions from food systems.

According to Sabaté and Soret (2014, p. 476S), adopting plant-based diets "is perhaps one of the most rational and moral paths for a sustainable future" that involves responding to the climate change challenge. As the remaining chapters of this book show, a switch to more plant-based foods is an important practical moral stance that can help the planetary emergency not only by reducing the greenhouse gas emissions, but also by decreasing the demand for land and land clearing, and resultant biodiversity loss, by preserving the marine environments as well as helping prevent diet-related non-communicable diseases.

We are writing this book in Australia, where the consumption of the animal-based products is one of the highest in the world. As at 27 August 2021, the animal kill clock (https://animalclock.org/au/) shows 3,235,460,867 animals slaughtered for food in Australia, including chickens, turkeys, cattle, pigs, ducks, sheep, fish and shellfish, since the start of the calendar year. This book is asking fellow Australians and any other readers to examine their food choices and take a moral stance against the wilful destruction of the planet. We are in a planetary emergency and the clock is ticking…

References

American Psychiatric Association (APA). (2021). *Climate change and mental health connections.* https://www.psychiatry.org/patients-families/climate-change-and-mental-health-connections

Bernabucci, U. (2019). Climate change: Impact on livestock and how can we adapt. *Animal Frontiers, 9*(1), 3–5. https://doi.org/10.1093/af/vfy039

Bishop, J. (2020). *Burnt assets: The 2019–2020 Australian bushfires.* WWF-Australia. https://www.wwf.org.au/news/news/2020/the-hidden-cost-of-australia-s-bushfire-crisis-could-be-billions#gs.0zwvwe

Bureau of Meteorology (BoM). (2021). *Annual climate statement 2020.* http://www.bom.gov.au/climate/current/annual/aus/

Carbon Brief. (2021). *Attributing extreme weather to climate change.* https://www.carbonbrief.org/mapped-how-climate-change-affects-extreme-weather-around-the-world

Centre for Disaster Philanthropy (CDP). (2020). *2019–2020 Australian bushfires.* https://disasterphilanthropy.org/disaster/2019-australian-wildfires/

Cho, R. (2018). *How climate change will alter our food.* Columbia Climate School. https://news.climate.columbia.edu/2018/07/25/climate-change-food-agriculture/

Cianconi, P., Betrò, S., & Janiri, L. (2020). The impact of climate change on mental health: A systematic descriptive review. *Frontiers in Psychiatry, 11.* https://doi.org/10.3389/fpsyt.2020.00074

Clark, M., & Tilman, D. (2017). Comparative analysis of environmental impacts of agricultural production systems, agricultural input efficiency, and food choice. *Environmental Research Letters, 12*(6), 064016. https://doi.org/10.1088/1748-9326/aa6cd5

Commonwealth Scientific and Industrial Research Organisation (CSIRO). (2021). *The 2019–20 bushfires: A CSIRO explainer.* https://www.csiro.au/en/research/natural-disasters/bushfires/2019-20-bushfires-explainer

Costello, C., Ovando, D., Clavelle, T., Strauss, C. K., Hilborn, R., Melnychuk, M. C., … Leland, A. (2016). Global fishery prospects under contrasting management regimes. *Proceedings of the National Academy of Sciences of the United States of America (PNAS), 113*, 5125–5129. https://doi.org/10.1073/pnas.1520420113

Department of Primary Industries and Regional Development (DPIRD). (2021). *Climate change and broadacre livestock production.* Government of Western Australia. https://www.agric.wa.gov.au/climate-change/climate-change-and-broadacre-livestock-production

Doctors for the Environment Australia (DEA). (2020). *Healthy planet, healthy people.* https://dea.org.au

Energy & Climate Intelligence Unit. (2021). *Net zero: Farming and the countryside.* https://eciu.net/analysis/briefings/net-zero/net-zero-farming-and-the-countryside

European Academies Science Advisory Council (EASAC). (2018). *Extreme weather events in Europe. Preparing for climate change adaptation: An update on EASAC's 2013 study.* https://easac.eu/fileadmin/PDF_s/reports_statements/Extreme_Weather/EASAC_Extreme_Weather_2018_web.pdf

Farmery, A. K., Hendrie, G. A., O'Kane, G., McManus, A., & Green, B. S. (2018). Sociodemographic variation in consumption patterns of sustainable and nutritious seafood in Australia. *Frontiers in Nutrition, 5*, 118. https://doi.org/10.3389/fnut.2018.00118

Food and Agriculture Organisation of the United Nations (FAO). (2016). *Livestock & climate change.* http://www.fao.org/3/i6345e/i6345e.pdf

Food and Agriculture Organisation of the United Nations (FAO). (2021). *Hazard and emergency types.* http://www.fao.org/emergencies/emergency-types/hazard-and-emergency-types/en/

Garnett, T. (2011). Where are the best opportunities for reducing greenhouse gas emissions in the food system (including the food chain)? *Food Policy, 36*(Suppl 1), S23–S32. https://doi.org/10.1016/j.foodpol.2010.10.010

Grace, P., & Barton, L. (2014, December 9). *Meet N2O, the greenhouse gas 300 times worse than CO2.* The Conversation. http://theconversation.com/meet-n2o-the-greenhouse-gas-300-times-worse-than-co2-35204

Hollowed, A. B., Barange, M., Beamish, R. J., Brander, K., Cochrane, K., Drinkwater, K., … Yamanaka, Y. (2013). Projected impacts of climate change on marine fish and fisheries. *ICES Journal of Marine Science, 70*(5), 1023–1037. https://doi.org/10.1093/icesjms/fst081

Hossain, N., & Green, D. (2011). *Living on a spike: How is the 2011 food price crisis affecting poor people?* Oxfam Research Reports. https://www-cdn.oxfam.org/s3fs-public/file_attachments/rr-living-on-a-spike-food-210611-summ-en_4.pdf

Huang, M., Ding, L., Wang, J., Ding, C., & Tao, J. (2021). The impacts of climate change on fish growth: A summary of conducted studies and current knowledge. *Ecological Indicators, 121*, 106976. https://doi.org/10.1016/j.ecolind.2020.106976

Intergovernmental Panel on Climate Change (IPCC). (2014a). *AR5 climate change 2014: Impacts, adaptation, and vulnerability.* Contribution of Working Group II to the Fifth Assessment Report of the Intergovernmental Panel on Climate Change. https://www.ipcc.ch/report/ar5/wg2/

Intergovernmental Panel on Climate Change (IPCC). (2014b). *Climate change 2014: Synthesis report.* Contribution of Working Groups I, II and III to the Fifth Assessment Report of the Intergovernmental Panel on Climate Change. https://www.ipcc.ch/report/ar5/syr/

Intergovernmental Panel on Climate Change (IPCC). (2019). *Climate change and land: An IPCC special report on climate change, desertification, land degradation, sustainable land management, food security, and greenhouse gas fluxes in terrestrial ecosystems.* https://www.ipcc.ch/srccl/

Intergovernmental Panel on Climate Change (IPCC). (2021). *Climate change 2021: The physical science basis*. Working Group I contribution to the Sixth Assessment Report of the Intergovernmental Panel on Climate Change. Cambridge University Press (forthcoming). https://www.ipcc.ch/report/sixth-assessment-report-working-group-i/

International Society Doctors for the Environment. (2016). *Mission statement*. https://www.isde.org

MacLeod, M. J., Hasan, M. R., Robb, D. H. F., & Mamun-Ur-Rashid, M. (2020). Quantifying greenhouse gas emissions from global aquaculture. *Nature Research, 10*, 11679. https://doi.org/10.1038/s41598-020-68231-8

Mariani, G., Cheung, W. W. L., Lyet, A., Sala, E., Mayorga, J., Velez, L., ... Mouillot, D. (2020). Let more big fish sink: Fisheries prevent blue carbon sequestration – Half in unprofitable areas. *Science Advances, 6*, eabb4848. https://doi.org/10.1126/sciadv.abb4848

Marinova, D., & Raphaely, T. (2018). Impact of vegetarian diets on the environment. In W. Craig (Ed.), *Vegetarian nutrition and wellness* (pp. 13–24). CRC Press/Taylor and Francis.

Myhre, G., Alterskjær, K., Stjern, C. W., Marelle, L., Samset, B. H., Sillmann, J., ... Stohl, A. (2019). Frequency of extreme precipitation increases extensively with event rareness under global warming. *Scientific Reports, 9*, 16063. https://doi.org/10.1038/s41598-019-52277-4

Padhy, S. K., Sarkar, S., Panigrahi, M., & Paul, S. (2015). Mental health effects of climate change. *Indian Journal of Occupational and Environmental Medicine, 19*(1), 3–7. https://doi.org/10.4103/0019-5278.156997

Parker, R. W. R., Blanchard, J. L., Gardner, C., Green, B. S., Hartmann, K., Tyedmers, P. H., & Watson, R. A. (2018). Fuel use and greenhouse gas emissions of world fisheries. *Nature Climate Change, 8*, 333–337. https://doi.org/10.1038/s41558-018-0117-x

Pathak, H., Jain, N., Bhatia, A., Patel, J., & Aggarwal, P. K. (2010). Carbon footprints of Indian food items. *Agriculture, Ecosystems & Environment, 139*(1–2), 66–73. https://doi.org/10.1016/j.agee.2010.07.002

Poore, J., & Nemecek, T. (2018). Reducing food's environmental impacts through producers and consumers. *Science, 360*(6392), 987–992. https://doi.org/10.1126/science.aaq0216

Ritchie, H., & Roser, M. (2021). *Environmental impacts of food production*. Our World in Data. https://ourworldindata.org/environmental-impacts-of-food

Rivera-Ferre, M. G., López-i-Gelats, F., Howden, M., Smith, P., Morton, J. F., & Herrero, M. (2016). Re-framing the climate change debate in the livestock sector: Mitigation and adaptation options. *Wiley's Interdisciplinary Reviews (WIREs) Climate Change, 7*(6), 869–892. https://doi.org/10.1002/wcc.421

Rojas-Downing, M., Pouyan Nejadhashemi, A., Harrigan, T., & Woznicki, S. A. (2017). Climate change and livestock: Impacts, adaptation, and mitigation. *Climate Risk Management, 16*, 145–163. https://doi.org/10.1016/j.crm.2017.02.001

Sabaté, J., & Soret, S. (2014). Sustainability of plant-based diets: Back to the future. *The American Journal of Clinical Nutrition, 100*(Suppl 1), 476S–482S. https://doi.org/10.3945/ajcn.113.071522

Sanderson, B. M., & Fisher, R. A. (2020). A fiery wake-up call for climate science. *Nature Climate Change, 54*, 175–177. https://doi.org/10.1038/s41558-020-0707-2

Scarborough, P., Appleby, P. N., Mizdrak, A., Briggs, A. D. M., Travis, R. C., Bradbury, K. E., & Key, T. J. (2014). Dietary greenhouse gas emissions of meat-eaters, fish-eaters, vegetarians and vegans in the UK. *Climatic Change, 125*(2), 179–192. https://doi.org/10.1007/s10584-014-1169-1

Scheelbeek, P. F. D., Bird, F. A., Tuomisto, H. L., Green, R., Harris, F. B., Joy, E. J. M., ... Dangour, A. D. (2018). Effect of environmental changes on vegetable and legume yields and nutritional quality. *Proceedings of the National Academy of Sciences of the United States of America (PNAS), 115*(26), 6804–6809. https://doi.org/10.1073/pnas.1800442115

Schmidinger, K., Bogueva, D., & Marinova, D. (2018). New meat without livestock. In D. Bogueva, D. Marinova, & T. Raphaely (Eds.), *Handbook of research on social marketing and its influence on animal origin food product consumption* (pp. 344–361). IGI Global.

Simpson, S., Blanchard, J., & Genner, M. (2013). Impacts of climate change on fish. *Marine Climate Change Impacts Partnership: Science Review, 2013,* 113–124. https://doi.org/10.14465/2013. arc13.113-124

Sippel, S., Meinshausen, N., Fischer, E. M., Székely, E., & Knutti, R. (2020). Climate change now detectable from any single day of weather at global scale. *Nature Climate Change, 10,* 35–41. https://doi.org/10.1038/s41558-019-0666-7

Soret, S., Mejia, A., Batech, M., Jaceldo-Siegel, K., Harwatt, H., & Sabaté, J. (2014). Climate change mitigation and health effects of varied dietary patterns in real-life settings throughout North America. *The American Journal of Clinical Nutrition, 100*(Suppl 1), 490S–495S. https://doi.org/10.3945/ajcn.113.071589

Tack, J., Lingenfelser, J., & Jagadish, S. V. K. (2017). Warming adaptation requires more genetic diversity. *Proceedings of the National Academy of Sciences of the United States of America (PNAS), 114*(35), 9296–9301. https://doi.org/10.1073/pnas.1706383114

Tadesse, G., Algieri, B., Kalkuhl, M., & von Braun, J. (2014). Drivers and triggers of international food price spikes and volatility. *Food Policy, 47,* 117–128. https://doi.org/10.1016/j.foodpol.2013.08.014

The Climate Institute. (2011). *A climate of suffering: The real cost of living with inaction on climate change.* The Climate Institute, Melbourne & Sydney, Australia. https://www.aph.gov.au/DocumentStore.ashx?id=20b0f0b2-775d-40d9-81ab-72aceafcdd8b

Tigchelaar, M., Battisti, D. S., Naylor, R. L., & Ray, D. K. (2018). Future warming increases probability of globally synchronized maize production shocks. *Proceedings of the National Academy of Sciences of the United States of America (PNAS), 115*(26), 6644–6649. https://doi.org/10.1073/pnas.1718031115

Tilman, D., & Clark, M. (2014). Global diets link environmental sustainability and human health. *Nature, 515,* 518–522. https://doi.org/10.1038/nature13959

Union of Concerned Scientists (UCS). (2020). *Each country's share of CO_2 emissions.* https://www.ucsusa.org/resources/each-countrys-share-co2-emissions

United Nations Office for Disaster Risk Reduction (UNODRR). (2020). *Human cost of disasters: An overview of the last twenty years 2000–2019.* https://www.undrr.org/publication/human-cost-disasters-2000-2019

United Nations Office for Disaster Risk Reduction (UNODRR). (2021). *2020: The non-COVID year in disasters. Global trends and perspectives.* https://www.undrr.org/publication/2020-non-covid-year-disasters

van Oldenborgh, G. J., Krikken, F., Lewis, S., Leach, N. J., Lehner, F., Saunders, K. R., … Otto, F. E. L. (2021). Attribution of the Australian bushfire risk to anthropogenic climate change. *Natural Hazards and Earth System Sciences, 21*(3), 941–960. https://doi.org/10.5194/nhess-21-941-2021

Wahlquist, C. (2020). *Up to 100,000 sheep killed in Kangaroo Island fires, as farmers tally livestock losses.* The Guardian. https://www.theguardian.com/australia-news/2020/jan/13/up-to-100000-sheep-killed-in-kangaroo-island-fires-as-farmers-tally-livestock-losses

Weatherdon, L. V., Magnan, A. K., Rogers, A. D., Sumaila, U. R., & Cheung, W. W. L. (2016). Observed and projected impacts of climate change on marine fisheries, aquaculture, coastal tourism, and human health: An update. *Frontiers in Marine Science, 3,* 48. https://doi.org/10.3389/fmars.2016.00048

Wedderburn-Bisshop, G., & Rickards, L. (2018). Livestock's near-term climate impact and mitigation policy implications. In D. Bogueva, D. Marinova, & T. Raphaely (Eds.), *Handbook of research on social marketing and its influence on animal origin food product consumption* (pp. 37–57). IGI Global.

Willett, W., Rockström, J., Loken, B., Springmann, M., Lang, T., Vermeulen, S., … Murray, C. J. L. (2019). Food in the Anthropocene: The EAT–Lancet Commission on healthy diets from sustainable food systems. *The Lancet, 393*(10170), 447–492. https://doi.org/10.1016/S0140-6736(18)31788-4

World Wide Fund for Nature (WWF) Australia. (2020). *Australia's 2019–2020 bushfires: The wildlife toll.* Interim Report. https://www.wwf.org.au/what-we-do/bushfire-recovery/in-depth/resources/australia-s-2019-2020-bushfires-the-wildlife-toll#gs.xuknob

Zhao, C., Liu, B., Piao, S., Wang, X., Lobell, D. B., Huang, Y., … Asseng, S. (2017). Temperature increase reduces global yields of major crops in four independent estimates. *Proceedings of the National Academy of Sciences of the United States of America (PNAS), 114*(35), 9326–9331. https://doi.org/10.1073/pnas.1701762114

Zhou, X., Passow, F. H., Rudek, J., Von Fisher, J. C., Hamburg, S. P., & Albertson, J. D. (2019). Estimation of methane emissions from the U.S. ammonia fertilizer industry using a mobile sensing approach. *Elementa: Science of the Anthropocene, 7*(1), 19. https://doi.org/10.1525/elementa.358

Chapter 3
Food and Environmental Emergency

Abstract In addition to climate change, the current planetary emergency is also triggered by significant anthropogenic modifications in the planet's biophysical systems. The chapter covers issues related to land use, water, soil fertility, application of chemical fertilisers and impacts on biodiversity. All these problems are interrelated and manifested differently in various parts of the world. The West on the one hand has been in many ways responsible for the current planetary emergency and on the other, is well-equipped to respond to it by changing its attitude and behaviour while adopting more plant-rich food choices with lower environmental impacts.

There are certain books that you remember for life and they also shape the way you see the world. One of them is Rachel Carson's *Silent Spring* originally published in 1962. This book has certainly impacted many people around the globe but for us it is very special as it was made available in Bulgaria during socialist times translated into Bulgarian. It was very rare to find a western book (in its original language or translated) in Bulgarian bookshops in the 1960s, 1970s and even in the 1980s and 1990s. Only classical master-pieces or exceptional books from the West would be allowed by the Communist Party to reach the broader Bulgarian public.

Silent Spring was such an exceptional book, a ground-breaking landmark in the foundation and development of the environmental movement. It was a plea for what we now call planetary health. The book exposed the consequences from big business using the chemical insecticide DDT – dichloro-diphenyl-trichloro-ethane with the chemical formula $(ClC_6H_4)_2CHCCl_3$ to increase yields. Chemically synthesised in 1874, the use of DDT as a pesticide was discovered in 1939 and during the Second World War, it was also found to be effective in the prevention of vector-borne diseases, such as malaria and typhus (International Programme on Chemical Safety, 1979). Dr. Paul Hermann Müller who was a chemist, worked as a laboratory technologist for the Swiss J. R. Geigy (a predecessor of Novartis), did not have any medical qualifications and never conducted medical research, received a Nobel prize in physiology or medicine in 1948 for the discovery of the insecticidal qualities of DDT in the control of insect-transmitted diseases (Berry-Caban, 2011).

D. Marinova, D. Bogueva, *Food in a Planetary Emergency*,
https://doi.org/10.1007/978-981-16-7707-6_3

Initially, this technological product was seen as being highly effective and was widely used in agriculture against pests. However, later on it was established that DDT is stable under most environmental conditions and does not break down completely by the microorganisms present in the soil or in the bodies of other organisms exposed to it (International Programme on Chemical Safety, 1979). The drift from vaporisation during field application of DDT was detected across the entire world; DDT accumulated in the soils and waters; residues were found in meat and dairy products and biomagnification occurred when humans consumed these foods. It also affected the human body where it stored in fatty tissues and during periods of starvation, DDT-related compounds were released in the blood system carried out by proteins (National Pesticide Information Centre, 1999). Furthermore, DDT became infamous for its negative ecological and health impacts on beneficial insects, such as bees and other pollinators; birds, where it caused thinning of the eggshells, other wildlife, including frogs and fish, and humans. For humans, DDT is carcinogenic, has serious impacts on the cardiovascular system, central nervous system, liver, causes tumours and mutations. To make things worse, some insects, such as the house fly, became resistant to DDT. The pollution of the natural environment through DDT application continued for many decades.

Bulgarian agriculture also used DDT; however, since the 1950s its toxicity was already known with official national security documents raising concerns about its effects on people (Agroclub.bg, 2019). Across the world, the scientific and empirical evidence about the undesirable effects of DDT was mounting. For example, in 1953, the journal *Nature* published a paper about the adverse changes in animal communities after the application of DDT (Sheals, 1953). High concentrations of DDT – up to 150 times the standard, were found in many Bulgarian rivers and dams whose waters were used for irrigation and drinking (Petel.bg, 2015). Consequently, the use of DDT was banned in Bulgaria in 1969, following the 1968 ban in socialist Hungary. Most western countries, including USA, banned the use of DDT in the 1970s with the exception of UK, where it continued to be used until 1984. The 2004 Stockholm Convention on Persistent Organic Pollutants (POPs) put a worldwide ban on the application of DDT for agricultural purposes (van den Berg et al., 2017).

There are many parallels that can be drawn between the story of DDT and other chemicals being applied nowadays to improve agricultural yields. The one that is being extremely controversial and getting a lot of negative publicity and legal attention is glyphosate widely used in USA since the 1970s and classified by the World Health Organisation (IARC & WHO, 2015) as probably carcinogenic (category 2A). It may take a further 50 years or more before a ban is imposed on such dangerous substances. One of the questions that *Silent Spring* posed was who benefits from such large-scale chemical applications in the pursue of higher yields and bigger profits. Under the "feeding the world" banner, chemical companies and large-scale agricultural producers have been allowed to cause substantial environmental harm contributing to the current planetary emergency. Profit-driven corporations and deceitful government institutions were allowed to exploit nature and cause enormous environmental damage by encouraging the production of single-crop yields.

In the words of Rachel Carson: "We have done it, moreover, for reasons that collapse the moment we examine them. We are told that the enormous and expanding use of pesticides is necessary to maintain farm production… Yet is our real problem not one of *overproduction*?" (Carson, 2000, p. 44).

Almost 60 years after *Silent* Spring, the scale of the food problem has significantly escalated and changed. The issue of overproduction in some parts of the world has also translated in overconsumption, particularly animal-based foods. It has worsened so much that it is now threatening human existence not only through the use of chemicals. The previous chapter explained the bidirectional relationship between food and climate change. This chapter is expanding the gamut of environmental concerns linked to our food choices. It is not our intention to tell the reader what to eat; however, it is important that people understand the facts around the food choices they make. We are adding our voices to Rachel Carson's plea for planetary health as it is not just DDT and chemical pesticides that are killing all species on the planet, including us. The problems around food production and consumption have become so complex and so intertwined with half-truths and half-science, that many people who are genuinely looking to make the right decisions are feeling lost. We are offering some insights. All species, including us, need to feed to survive and this book explains that by making better food choices we can respond to the current planetary emergency.

The aspects that we cover in this chapter relate to land use, water, soil fertility, application of chemical fertilisers and biodiversity. Food packaging discussed later in the book is also significantly contributing to the current serious environmental problems. All these problems are interrelated and manifested variously in different parts of the world. We cannot escape but offer a very much western perspective. However, the West has been in many ways responsible for the current planetary emergency and also is well-equipped to respond by changing its attitude and behaviour towards food.

Land Use

Before the industrial revolution, most of the land mass available on Earth was wilderness with wild species of plants and animals. One thousand years ago, only 4% of the world's ice-free and non-barren land was used for farming (Ritchie & Roser, 2019). This has radically changed and by now half of the ice-free land is used by agriculture (Ellis et al., 2010). Land conversion from wilderness to croplands, pastures and food-growing lots has had enormous impacts on the health of the planet with hundreds of fauna and flora species lost, depletion of soils and release of greenhouse gas emissions. With industrialisation, cities have attracted people in their pursuit of education, employment and better quality of life. Cities are now home to 56.2% of the global population (Satterthwaite, 2020). Urban dwellers often do not see the scale of the enormous land changes as they are happening outside of the built-up areas that take up only 1% of the total planet's land mass (see Fig. 3.1).

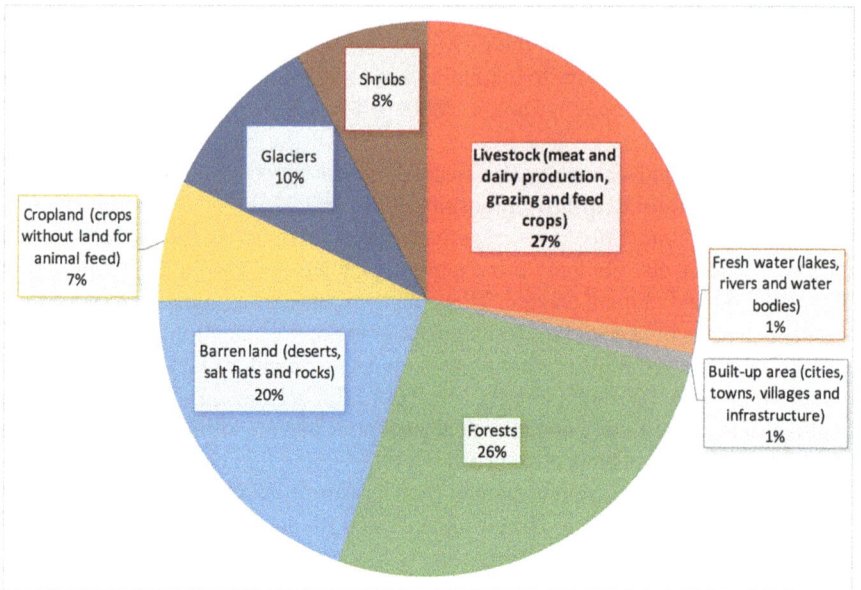

Source of data: Ritchie & Roser (2019)

Fig. 3.1 Global land use (percentages). (Source of data: Ritchie & Roser, 2019)

Another change that has become unnoticed is the expansion of the livestock sector over agricultural and other land. More than two-thirds (namely 77%) of the available agricultural land on the planet are now being used for livestock grazing and feed to supply only just above a third (namely 37%) of the global protein (Ritchie & Roser, 2019). Clearing of native vegetation and conversion of vast areas into pastures or crop fields to grow animal feed, such as corn and soybeans, are happening all across the globe. Each year, 13 billion hectares (or 130 million km^2) of forest and native vegetation are lost globally because of the livestock sector (FAO, 2012). Below are a couple of examples from countries where meat consumption and related exports thrive. Over 50% of the Cerrado – a vast tropical savannah region located at the heart of Brazil covering 1.9 million km^2, has been cleared for pastures and animal crops (Spera et al., 2020). More than 90% of the 24.5 thousand km^2 cleared land in the state of Queensland in Australia between 2010 and 2018 has been for pastures (Kilvert, 2020). The livestock sector is in fact the largest anthropogenic user of land (Marinova & Raphaely, 2018).

Land requirements for the diets across the world largely differ. An estimate by Robbins (2012) shows that the standard American (or Western) diet requires 3 acres (or 12 thousand m^2) per person or 18 times more compared to the 1/6 acres (or 675 m^2) needed for a plant-only (or vegan) diet and 6 times more than the 1/2 acres (or 2 thousand m^2) necessary for a vegetarian diet (which includes eggs and dairy). Organic farming without the use of fertilisers, pesticides and genetically modified organisms (GMOs) is considered to be better for the health of the planet; however, it requires even larger land areas because of lower yields. According to Treu et al. (2017), an organic diet which contains 45% less meat than the standard German diet, still requires 40% more land than the conventional diet. A meta-analysis of 742 agricultural systems and over 90 foods (Clark & Tilman, 2017) similarly shows that organic systems require more land and cause more eutrophication than conventional systems and although use less energy, they emit similar greenhouse gas emissions. Hence, growing organic food cannot be the solution to feeding the current 7.9 billion population on an omnivorous diet, but a change towards foods with less environmental impact can be.

A map of the world produced by Our World in Data (see Fig. 3.2) shows the countries that are currently exceeding the global land constraints (in orange) if the total world population is to adopt their diet. This includes the diet of New Zealand requiring almost double the global habitable land of the planet, Argentina – 1.6 times, Australia – 1.5 times, USA – 1.4 times, Canada – 1.3 times and Brazil – 1.2 times. Adopting the diets of the countries in yellow would require expansion of the current agricultural land at the expense of the remaining wild areas but cannot theoretically be achieved. Only the diets of the countries in blue are the ones that can allow humanity to cut its use of agricultural land. However, the empirical evidence shows that the richer the population is or becomes, the higher its meat consumption and land demand are (Ritchie, 2017).

Share of global habitable land needed for agriculture if everyone had the diet of...
The percentage of global habitable land area needed for agriculture if the total world population was to adopt the average diet of any given country, in 2011. The actual proportion of habitable land used for agriculture was 50 percent. Values greater than 100% are not possible within global land constraints.

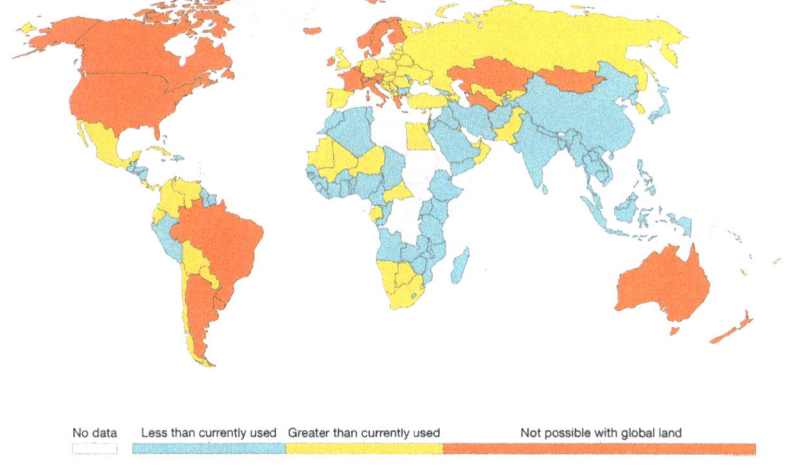

Source: Our World in Data (https://ourworldindata.org/agricultural-land-by-global-diets)

Fig. 3.2 Land needed if the world adopts a particular country's diet. (Source: Our World in Data. https://ourworldindata.org/agricultural-land-by-global-diets)

In order to feed the world, we need changes not only in the behaviour of the rich sections of the global population but also in what kind and in which ways food is produced. We expand on these issues later in the book. For the time being, however, we want to stress that it is not just doom and gloom and that the world can indeed feed its human population. According to data from the World Bank (2021), in 2018 the global agricultural land was 48 million km^2 and the global agricultural land (where crops can be produced) was 14 million km^2. This arable land can support a global human population of 11.7 billion on a vegetarian diet and 20 billion on a plant-based diet. It cannot however support the current global population on an American or Australian diet. Shifting to more plant-based choices is good not only to feed the global population but also for human health and for the well-being of the planet's natural environment.

Soil Fertility

Soil is the foundation of agriculture and the majority of food production on Earth. Its properties determine the ability to grow plants. The natural cycling of the 17 essential nutrient elements for the growth of crops (see Table 3.1) occurs through plants to animals, including people, then back into the soil as a result from the biological processes of decomposition (Parikh & James, 2012). By comparison, although the functional nutrients play an important role in the development of

Table 3.1 Essential and functional nutrient elements for plants

Essential non-mineral elements and symbols	Essential mineral elements and symbols	Functional mineral elements and symbols
Carbon (C)	*Primary macronutrients*	Sodium (Na)
Hydrogen (H)	Nitrogen (N)	Selenium (Se)
Oxygen (O)	Phosphorus (P)	Silicon (Si)
	Potassium (K)	
	Secondary macronutrients	
	Calcium (Ca)	
	Magnesium (Mg)	
	Sulphur (S)	
	Micronutrients	
	Boron (B)	
	Chlorine (Cl)	
	Copper (Cu)	
	Iron (Fe)	
	Manganese (Mn)	
	Molybdenum (Mo)	
	Nickel (Ni)	
	Zinc (Zn)	

Source: Troeh and Thompson (2005) and Subbarao et al. (2003)

plants, including nutrition, metabolism, growth, strength and resistance to fungal diseases, they are widely available in the soils and do not depend on the biological processes of decomposition. Soil organic matter, which comprises of partially or fully-decomposed residues in the soil, is particularly valuable for agriculture (Parikh & James, 2012). However, agriculture, and specifically intensive agricultural methods, significantly alter the normal cycling of essential nutrients with crops essentially sucking them out of the soil resulting in a reduced, and in some cases depleted, fertility. Globally between 1 and 6 billion hectares (10–60 million km^2) of land is already degraded due to agricultural activities (Gibbs & Salmon, 2015). Applying synthetic fertilisers is a way to compensate for the reduced soil fertility but ultimately adds to the threats of climate change and biodiversity loss.

The increasing demand for production of larger quantities of crops for human and animal consumption puts additional pressure on the soils disrupting even further the natural cycling of nutrients. Soils also act as both, a sink and source for carbon, and there are many suggestions how to mitigate greenhouse gas emissions to improve their storage ability, including crop rotation, reduced tillage, agroforestry and reversing croplands to a land cover similar to native vegetation (Smith et al., 2008). Decreased moisture in soils as a result from temperature increases affects the biological decomposition rates of organic matter and slows even further the natural cycling of nutrients (Smith, 2012). On the other hand, the increased plant growth expected with higher levels of CO_2 in the atmosphere requires plants to extract nutrients from the soil even faster decreasing its ability to store carbon (Terrer et al., 2021).

Currently, soils store three times the amount of carbon in vegetation and about twice the amount of carbon present as carbon dioxide in the Earth's atmosphere (Smith, 2012). However, the rate of soil organic matter formation and regeneration is very slow. It takes between 500 and 1000 years for one inch of soil organic matter to form and agricultural practices since the industrial revolution have contributed to massive soil erosion (Cho, 2012). According to Parikh and James (2012), soils should be considered a non-renewable resource that needs to be preserved and used with care. This requires us to re-examine the current use of agricultural soils which are found only on 37% of the Earth's land surface (FAOSTAT, 2021). As the statistics of the UN Food and Agriculture Organisation (FAOSTAT, 2021) show, since 1997 we have lost 1% of global land suitable for agricultural purposes and during this time population numbers have grown from 5.9 billion to 7.9 billion. Fertiliser use has also significantly increased, including the application of nitrogen (N) responsible for the potent greenhouse gas nitrous oxide (N_2O) – from 83 million to 109 million tonnes. We are facing a serious problem with soil fertility about which there is not enough awareness or action to prevent its escalation.

Livestock and the consumption of animal-based foods contribute disproportionately more to soil depletion than plant-based foods in several major ways. The first is through overgrazing which removes vegetation from the soils exposing them to erosion by winds, precipitations and surface run-off. This grazing often occurs in areas that have been cleared from native vegetation, such as rainforests, steppes, savannahs and bush (in the case of Australia), changing the composition of the soils and increasing salinity. In addition, the crops grown for animal feed further extract nutrients from the soils to provide protein for human consumption in a very inefficient way. In the case of beef, it takes 38 calories fed to the animal to produce one calorie for human consumption (Eshel et al., 2014). In essence, we are gambling with the future of the planet's soils by continuing to raise large numbers of livestock animals. If increasing carbon sequestration in soils is seen as one of the pathways to achieve the Paris Agreement (Zomer et al., 2017), the way this precious resource is being used needs to change.

Chemical Fertilisers

At the time of *Silent Spring* and afterwards, life scientists and chemical engineers were trying to find ways to increase food production yields conquering pests, diseases and environmental challenges. As global climate change now is destabilising many natural processes of growing food, these specialists continue to be constantly on the look for technological solutions. Trusting the power of innovation to improve yields, they persist in pursuing the same goal of improving food quality and quantity under the changing climatic and soil fertility conditions. Chemical, artificial, mineral or synthetic fertilisers have been part of the answer.

The era of artificial fertilisers was born with the invention of the German chemist Fritz Haber who created the Haber-Bosch process, a way of converting nitrogen from the air into liquid ammonia (NH_3) described as "providing bread from air" (Dunikowska & Turko, 2011, p. 10052). He was awarded the Nobel Prize in Chemistry in 1918 for his world-changing discovery. Fritz Haber also pioneered chemical warfare and the use of deadly poison gases in the battlefields of the First World War. Although ammonia per se is short-lived in the atmosphere and is not a greenhouse gas, the Haber-Bosch process which continues to be widely used for its production, is energy-intensive and contributes to the carbon footprint of agriculture (Vince, 2012).

Intensified farming after the Second World War, the spread of the agricultural Green Revolution in Asia and South America in the 1960s (World Food Summit, 1996) and many present-day cultivating practices heavily rely on the application of synthetic fertilisers containing nitrogen and phosphorus (National Geographic, 2020) to achieve high productivity of crops in high demand, such as corn, wheat, rice and other grains. The human-made combinations of nitrogen, phosphorus, potassium, calcium, magnesium and other chemical elements and inorganic substances are the basis of synthetic fertilisers, which immediately supply these essential nutrients to the soil unlike their organic counterparts that require longer periods of decomposition. A long-term negative effect however is the killing of the soil's beneficial microorganisms which naturally help convert dead animal and plant material into valuable and nutrient-rich organic matter. Also, in the long run, plants grown in over-fertilised soils, which stimulate fast growth, become deficient in crucial nutritional value lacking minerals and vitamins, such as iron, zinc, carotene, vitamin C, copper and protein.

In addition to ammonia, nitrates and urea are other nitrogen-based fertilisers widely used in agriculture. Globally, China was the largest producer of nitrogen fertilisers (36 million tonnes) in 2019, followed by India (14 million tonnes) and the United States (11 million tonnes) (Nation Master, n.d.). In 2019, China also produced phosphate and potash (18 million tonnes) as well as imported (11 million tonnes) fertilisers to boost its soil productivity (Wong, 2021). The use of nitrogen fertilisers is creating a global environmental and difficult to handle problem. Although nitrogen from synthetic fertilisers is readily available for plants to absorb (compared to that from organic fertilisers which needs to be converted first to inorganic), only half of all applied nitrogen is taken up by crops (Millar, 2015). If fertiliser nitrogen is not taken up by plants, it is hard to be contained and spreads in groundwaters as nitrate or in the air as nitrous oxide (N_2O) – a highly potent greenhouse gas, with a global warming potential 265–298 times that of CO_2 and a lifespan of more than 100 years (US EPA, 2020). In USA, for example, agriculture contributes 75% of all N_2O emissions linked to human activities (Millar, 2015).

Nitrogen is present in the Earth's atmosphere as an inert gas and is a critical ingredient for the existence of life. For most of the history of the planet, it has been available only to plants and animals on a limited scale through nitrogen-fixing bacteria and algae. By adding readily available synthetic fertiliser nitrogen to the planet's soils, we are allowing nitrogen to penetrate groundwaters and other water sources causing long-term damage and disturbing the whole ecosystem as well as to dangerously accumulate in the atmosphere and contribute to climate change.

The use of phosphorous for over 50 years to produce phosphate fertilisers and boost crop yields has resulted in its exploitation at a faster rate than nature can replenish it (Cordell & White, 2014). We are running out of this non-renewable resource having depleted the available phosphate rocks and crossed the planetary boundary in this area (see Introduction). Phosphorous has no substitute for soil health and in food production. The only response we have left is to transition to better, more sustainable food production and consumption practices that allow beneficial microorganisms to colonise the soils. In the chapter about circular agriculture, we present possible solutions, such as leaving roots, stems and leaves in the ground, making environmental nutrients available to plants, fixing nitrogen from the air and breaking down inorganic phosphates. Ultimately, the aim is to minimise the use of chemical fertilisers (Bomgardner, 2020).

Water

Freshwater is the foundation of life on this planet and a major consideration when growing food. Although 70% of the globe is covered by water, only 2.5% of it is freshwater (National Geographic, n.d.), including glaciers which contain almost 70% of all world's freshwater and are severely threatened by climate change (NSIDC, 2020). We use freshwater at a much faster rate than population growth; for example, between 1941 and 2011 the world's population tripled but its water consumption quadrupled (Oppenlander, 2013). Despite some technological advances, such as desalination, agriculture largely depends on freshwater. Agriculture is the main user of freshwater responsible for 69% of all global use each year (Piesse, 2020). Estimates show that livestock agriculture uses between 20% and 33% of the global freshwater (Mekonnen & Hoekstra, 2012) but in some countries, such as USA, the share of livestock agriculture in total freshwater use is much higher at 55% (Jacobson et al., 2006).

Water is not a fully renewable resource, mainly because of overconsumption. It is also a limited resource with only 1% of the world's surface representing freshwater (Ritchie & Roser, 2019). Disputes between downstream and upstream users are common not only between nations (e.g. between India and Bangladesh related to the water from the Ganges River), but also between different parts of the same country (e.g. among the Australian states along the Murray-Darling Basin). In a water-constrained world, such disputes can easily escalate into serious political and military conflicts because of the fundamental importance of this natural resource. With freshwater scarcity, also come poverty and hunger, as often seen in Africa and parts of Asia, adding essential dimensions to the complexity of the problem.

We use surface freshwater from rivers, lakes, other waterbodies, dams, barrages, reservoirs and rainwater collection, but also draw groundwater from aquifers through wells or pumps. Water is further stored as soil moisture and in the Earth's atmosphere. Snow and ice coverage are other extensive areas of freshwater. The availability of freshwater, however, is changing because of unsustainable groundwater extraction, climate change or the combination of these two factors (Rodell et al., 2018) with dry areas becoming drier and wet areas in the tropics and high latitudes becoming wetter. Water scarcity is threatening many parts of the world with the projections for the Middle East–North African region particularly alarming.

Nearly half of the water used in agriculture comes from aquifers (Famiglietti & Ferguson, 2021) and 70% of the withdrawn groundwater is used in agriculture (Margat & van der Gun, 2013). Groundwater is the most extracted raw material globally at a rate of 982 km^3/year (Margat & van der Gun, 2013) and we are fast depleting this source which took tens of thousands of years to form. Alarmingly, more than half of the world's aquifers are being depleted (Famiglietti & Ferguson, 2021) with 6–20% of the wells (between 3 and 8 million across 40 countries) at risk of running dry if groundwater levels decline by 5 meters (Jasechko & Perrone, 2021). Many freshwater aquifers are at risk of saltwater intrusion, particularly in coastal areas, and of salination in arid and semi-arid areas. Continuous extraction of groundwater can also cause subsidence, permanently restricting the recharge and use of the aquifer.

Agriculture, including the livestock industry, is also responsible for polluting the waterways with untreated fertiliser runoffs and antibiotics which have not been metabolised in the animal bodies. A substantial part of this pollution is nitrate compounds which contribute to algal blooms (including the toxic cyanobacteria) and dead zones in the waterways as well as pose threats to human health, such as methemoglobinemia (known as blue baby syndrome), caused by drinking water with high levels of nitrates (Galan, 2018). Tyson Foods is USA's second polluter based on the company's reported toxic pollutants to the Environmental Protection Agency (EPA), which do not include pollution from factory farms raising livestock for the food giant (Environment America Research & Policy Center, n.d.). A third of the waterbodies in USA are polluted by agricultural runoff (Von Alt, 2016) while almost all rivers in New Zealand are contaminated with 60% of them above acceptable levels (Melhem, 2021).

Antibiotics are commonly used to prevent the spread of diseases among livestock animals living in crowded conditions, particularly poultry. As up to 90% of the anti-biotic substances are not properly metabolised, they are excreted through the urine and faeces (Bartlett et al., 2013) entering the water bodies and contributing towards building antimicrobial resistance (we discuss the issue of antibiotic use in further details later in the book). If all pollutants are included in assessing the environmental impacts of all businesses, the livestock sector would be by far the largest polluter of freshwater waterbodies.

It is hence very important how we use this precious resource which we share with hundreds of billions of other animals, including 81 billion livestock animals we currently raise for meat and milk (FAO, 2021). New technologies and practices, such as desalination, drip irrigation, extracting water from vapours in the atmosphere, managed aquifer recharge and water recycling, can ease the severity of the problem but cannot provide a complete solution. Some of these technologies come with the added challenges related to energy use and associated emissions with renewable sources becoming essential, as well as disposal of the saline brine which can add to the salination problems on the land and water potentially affecting other species. The main response to the looming global freshwater crises is in the foods we eat and in exposing the related water-intensive practices that threaten the availability of this precious natural resource (Oppenlander, 2013). Campaigns aimed at changing people's behaviour related to taking shorter showers, using water-efficient laundry machines and dishwashers to save water, pale in insignificance compared to addressing the problems of animal-based food consumption.

The quantity of water used to produce 1 kg of food varies between geographic locations and applied agricultural methods. It is clear that animal-based foods use much more water resources than crops with the exception of nuts and to a certain extent, rice (see Fig. 3.3). The data on Fig. 3.3 are based on the assessments by Poore and Nemecek (2018); however estimation methods vary. For example, Chapagain and Hoekstra (2004) estimate the water content of beef at 13,193 l/kg in USA, 17,122 l/kg in Australia and a world average at 15,497 l/kg which is significantly higher than the values on Fig. 3.3. Not all nuts have the same high-water requirements. For example, macadamia nuts use 1176 l/kg (The Macadamia, 2021), which is 3.5 times less water than the number shown on Fig. 3.3, with the non-irrigated trees producing the best yields.

As food is consumed to give energy and nutrition, many estimates also examine water use per calorie or gram of protein. Assessments again vary. For example, Pimentel and Pimentel (2003) estimate that the production of 1 kg of animal protein in USA requires about 100 times more water than 1 kg of grain protein. According to the study by Eshel et al. (2014), per unit of calorie, beef in the American diet requires respectively 4, 8 and 40 times more irrigated water than rice, potato and wheat. Figures 3.4 and 3.5 show the estimates of water use per 100 g of protein and

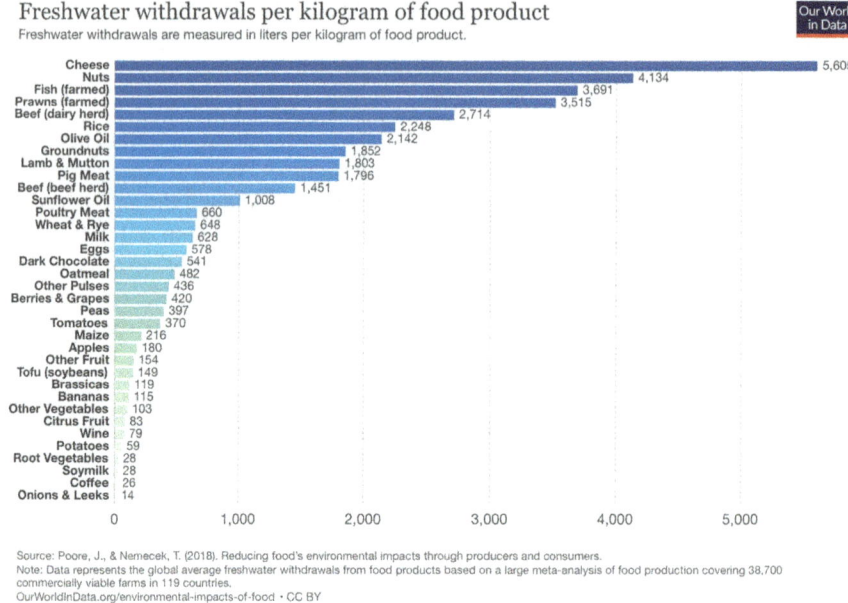

Fig. 3.3 Freshwater use per kilogram of food product. (Source: Ritchie & Roser, 2020)

1000 kilocalories based on data from Poore and Nemecek (2018). With future improvements in our understanding of the water types, cycles and use in growing and producing food, we are likely to see better consistency in the assessment of their water requirements. Notwithstanding the current differences in assessments, a reduction in the consumption of animal-based products replacing them with grains and vegetables will be beneficial in responding to the water scarcity on the planet. This is a much more important behaviour change that people need to make than eating locally or reducing the length of the showers they take. As the Australian Less Meat Less Heat website (2017) explains, eating one burger has the same water footprint as a month of daily showers.

Biodiversity

Deforestation and removal of native vegetation are the primary reason behind habitat destruction, reduction of wilderness and biodiversity loss. Half of the habitable land on the planet is already converted for agricultural purposes (Ritchie & Roser, 2019). Land fragmentation is another factor affecting biodiversity. Only 20% of the tropical forests are larger than 500 km² (Potapov et al., 2017) and globally 70% of the forests are within 1 km to a forest edge (Haddad et al., 2015). Livestock

Freshwater withdrawals per 100 grams of protein
Freshwater withdrawals are measured in liters per 100 grams of protein.

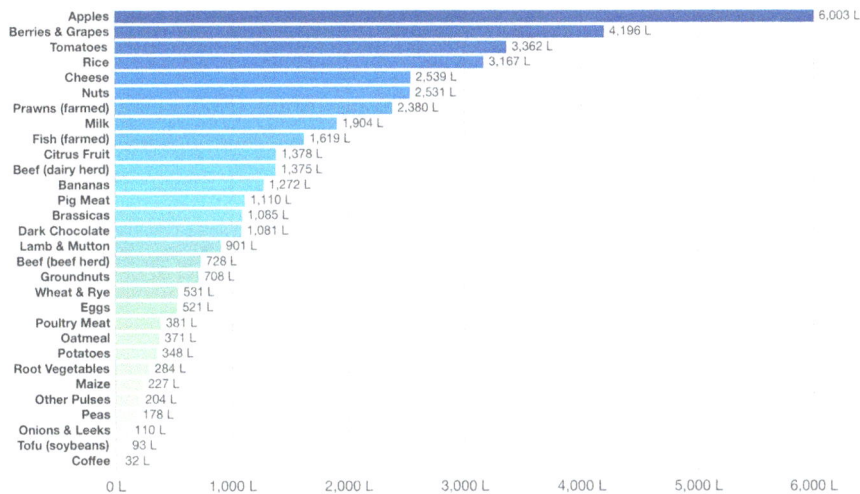

Source: Poore, J., & Nemecek, T. (2018) Additional calculations by Our World in Data   OurWorldInData.org/environmental-impacts-of-food • CC BY
Note: Data represents the global average freshwater withdrawals of food products based on a large meta-analysis of food production covering 38,700 commercially viable farms in 119 countries.

Source: Ritchie & Roser, 2020.

Fig. 3.4 Freshwater use per 100 grams of protein in food product. (Source: Ritchie & Roser, 2020)

agriculture has been the main driver for land-use changes and is behind 80% of deforestation and loss of wilderness worldwide (Kissinger et al., 2012). In the Australian state of Queensland, the livestock industry is responsible for 90% of the land clearing of native vegetation (Owens & Robinson, 2019).

Biodiversity defined as "diversity within species, between species and of ecosystems – is declining faster than at any time in human history" (IPBES, 2019, p. 10). The priority given to food production, including on the land and in the oceans, has occurred at the cost of many ecological services provided by nature, including fresh air, clean water and climate regulation. While agricultural output has been on the rise since the 1970s, the majority of life-regulating and non-material ecoservices of the natural environment have diminished with the abundance of native species declining by at least 20% in most major terrestrial biomes (IPBES, 2019). The impacts on the oceans are less understood but 66% of the marine environments are experiencing increasing cumulative impacts (IPBES, 2019) from pollution associated with food production, including plastics and discharge of sewerage from intensive farms and slaughterhouses. According to the assessment of the Intergovernmental Science-Based Platform on Biodiversity and Ecosystem Services (IPBES, 2019), 25% of animal and plant species are threatened translating to 1 million species facing extinction. Agriculture and aquaculture are listed as a threat for 36,510 species (or 98%) out of the 37,400 species threatened with extinction included in the Red List of the International Union for Conservation of Nature (IUCN, 2021). The

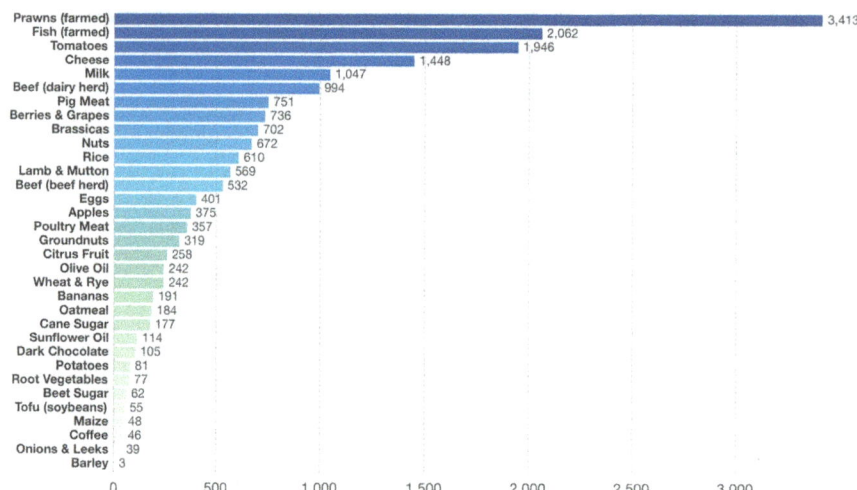

Source: Ritchie & Roser, 2020.

Fig. 3.5 Freshwater use per 1000 kilocalories of food product. (Source: Ritchie & Roser, 2020)

extinction rates amongst insects, which constitute the largest group of animals on the planet and provide critical ecoservices, including pollination, are eight times higher than for mammals, birds and reptiles, and are likely to plunge nature into a catastrophic collapse within a century (Sánchez-Bayo & Wyckhuys, 2019). Their dramatic rates of decline are due to habitat loss and land conversion to intensive agriculture with mono-cultures, use of synthetic pesticides and fertilisers as well as climate change (Sánchez-Bayo & Wyckhuys, 2019). A rethinking of how we produce food and the ecological price of animal-based proteins is long overdue.

The biomass of farmed animals, mainly cattle and pigs – namely 0.1 Giga tonnes of carbon (Gt C), now exceeds that of humans (namely 0.06 Gt C) by 67% (Bar-On et al., 2018). Most importantly, the combined biomass of livestock and humans now accounts for 96% of all mammals on the planet and is 53 times higher than the biomass of the wild land mammals (estimated at 0.003 Gt C) completely dominating the animal species (Bar-On et al., 2018). This is having profound and far-reaching ecological effects. The current agricultural and livestock practices are also the main reason for the loss of genetic biodiversity not only in the wild but also among the domesticated species of plants and animals. We have already lost close to 10% of all domesticated mammals and at least 1000 more are threatened with extinction (IPBES, 2019). Large-scale and broad-acre operations are based on the exploitation, in fact over-exploitation, of a limited number of species which makes them susceptible to pathogens and pest invasion. This was an issue Rachel Carson raised in 1962 and it has since become worse. The tentacles of human food production are spreading across the land and the oceans leaving behind devastation that ultimately threatens our own human existence. Many traditional and indigenous ways of producing food appear to be long-forgotten or are at least not embraced by industrial agriculture.

In 2010, the United Nations adopted the 20 Aichi Biodiversity Targets to be achieved during the 2011–2020 Decade of Biodiversity, aimed at addressing the main causes of biodiversity loss, reducing the pressure on and safeguarding the world's ecosystems (Convention on Biological Diversity, n.d.). None of these targets were achieved at a global level by 2020 and when there has been some progress, it relates mainly to policy arrangements, such as inclusion of biodiversity values in national accounting systems and expansion of protected areas (Secretariate of the CBD, 2020). No significant progress has been made in relation to the actual impacts of agriculture. Deforestation persists, be it at a slower rate, and agriculture continues to be the main driver of biodiversity loss. The health of the planet's ecosystems continues to decline and "mammal and bird species responsible for pollination are on average moving closer to extinction" (Secretariate of the CBD, 2020, p. 10). We are facing a silent spring.

Conclusion

Is everything lost or is it possible to turn the corner? The Sustainable Development Goals (SDGs) have taken over some of the tasks which the global community could not previously achieve, including those related to food production. Will they be achieved? What changes are needed?

We often return to Rachel Carson's book when reflecting about the enormity of the food problem: "One important natural check is a limit on the suitable habitat for each species" (Carson, 2000, p. 452). As at 12 pm on 22 May 2021, the size of the human species is 7,867,348,925 (Worldometers, 2021) – 7.9 billion is the largest human population number that this planet has seen. However, the numbers of livestock that this human population is killing each year to feed itself are staggering – a total of 80 billion sentient beings slaughtered annually, including 325 million cattle, 1.3 billion pigs and 72 billion chickens (FAO, 2021). There are many signs that human population numbers are going to stabilise by 2100 (Cillufo & Ruiz, 2019), but we also need to drastically decrease the size of the food-destined animals. Only then will we be able to have enough suitable habitat for each species, including wild animals and insects, native vegetation and agricultural crops. It is time to end the "era dominated by industry" – be it livestock or chemical industry, "in which the right to make a dollar at whatever cost is seldom challenged... The public must decide whether it wishes to continue on the present road, and it can do so only when in full possession of the facts" (Carson, 2000, p. 47). This book is making a humble contribution in laying out the facts about what the current situation is and what our alternatives are.

References

Agroclub.bg. (2019). *Spraying with the carcinogenic herbicide glyphosate resembles the historical DDT madness*. https://agroclub.bg/news/пръскането-с-канцерогенния-хербицид-глифозат-наподобява-безумието-с-ддт-1-504 [in Bulgarian].

Bar-On, Y. M., Phillips, R., & Ron Milo, R. (2018). The biomass distribution on Earth. *Proceedings of the National Academy of Sciences of the United States of America (PNAS), 115*(25), 6506–6511. https://doi.org/10.1073/pnas.1711842115

Bartlett, J. G., Gilbert, D. N., & Spellberg, B. (2013). Seven ways to preserve the miracle of anti-biotics. *Clinical Infectious Diseases, 56*(10), 1445–1550. https://doi.org/10.1093/cid/cit070

Berry-Cabán, C. (2011). DDT and *Silent Spring:* Fifty years after. *Journal of Military and Veterans' Health, 19*(4), 20–25.

Bomgardner, M. M. (2020). Crop innovations can protect yields and improve food quality in a changing climate. *Chemical & Engineering News, 98*(6). https://cen.acs.org/food/agriculture/Protecting-harvest/98/i6

Carson, R. (2000). *Silent spring*. The Folio Society.

Chapagain, A. K., & Hoekstra. A. Y. (2004). *Water footprints of nations* (Volume 1: Main report). UNESCO-IHE, Institute for Water Education. https://waterfootprint.org/media/downloads/Hoekstra_and_Chapagain_2006.pdf

Cho, R. (2012). *Why soil matters*. Columbia Climate School. https://news.climate.columbia.edu/2012/04/12/why-soil-matters/

Cillufo, A., & Ruiz, N. G. (2019). *World's population is projected to nearly stop growing by the end of the century*. Pew Research Centre. https://www.pewresearch.org/fact-tank/2019/06/17/worlds-population-is-projected-to-nearly-stop-growing-by-the-end-of-the-century/

Clark, M., & Tilman, D. (2017). Comparative analysis of environmental impacts of agricultural production systems, agricultural input efficiency, and food choice. *Environmental Research Letters, 12*, 064016. https://doi.org/10.1088/1748-9326/aa6cd5

Convention on Biological Diversity. (n.d.). *Aichi biodiversity targets*. https://www.cbd.int/sp/targets/

Cordell, D., & White, S. (2014). Life's bottleneck: Sustaining the world's phosphorus for a food secure future. *Annual Review of Environment and Resources, 39*, 161–188. https://doi.org/10.1146/annurev-environ-010213-113300

Dunikowska, M., & Turko, L. (2011). Fritz Haber: The damned scientist. *Angewandte Chemie International Edition, 50*, 10050–10062. https://doi.org/10.1002/anie.201105425

Ellis, E. C., Klein Goldewijk, K., Siebert, S., Lightman, D., & Ramankutty, N. (2010). Anthropogenic transformation of the biomes, 1700 to 2000. *Global Ecology and Biogeography, 19*(5), 589–606. https://doi.org/10.1111/j.1466-8238.2010.00540.x

Environment America Research & Policy Center. (n.d.). *America's next big polluter: Corporate agribusiness*. Company profile Tyson Foods, Inc. https://environmentamerica.org/sites/environment/files/reports/Env_Am_Tyson_v4.pdf

Eshel, G., Shepon, A., Makov, T., & Milo, R. (2014). Land, irrigation water, greenhouse gas, and reactive nitrogen burdens of meat, eggs, and dairy production in the United States. *Proceedings of the National Academy of Sciences of the United States of America (PNAS), 111*(33), 11996–12001. https://doi.org/10.1073/pnas.1402183111

Famiglietti, J. S., & Ferguson, G. (2021). The hidden crisis beneath our feet. *Science, 372*(6540), 344–345. https://doi.org/10.1126/science.abh2867

FAOSTAT. (2021). *World*. http://faostat.fao.org/static/syb/syb_5000.pdf

Food and Agriculture Organisation of the United Nations (FAO). (2012). *Livestock and land-scapes*. http://www.fao.org/3/ar591e/ar591e.pdf

Food and Agriculture Organisation of the United Nations (FAO). (2021). *FAOSTAT: Livestock primary*. http://www.fao.org/faostat/en/#data/QL

Galan, N. (2018). *What is blue baby syndrome?* Medical News Today. https://www.medicalnews-today.com/articles/321955

Gibbs, H. K., & Salmon, J. M. (2015). Mapping the world's degraded lands. *Applied Geography, 57*, 12–21. https://doi.org/10.1016/j.apgeog.2014.11.024

Haddad, N. M., Brudvig, L. A., Clobert, J., Davies, K. F., Gonzalez, A., Holt, R. D., … Townshend, J. R. (2015). Habitat fragmentation and its lasting impact on Earth's ecosystems. *Science Advances, 1*(2), e1500052. https://doi.org/10.1126/sciadv.1500052

Intergovernmental Science-Based Platform on Biodiversity and Ecosystem Services (IPBES). (2019). *Summary for policymakers of the global assessment report on biodiversity and ecosystem services of the Intergovernmental Science-Policy Platform on Biodiversity and Ecosystem Services* (S. Díaz, J. Settele, E. S. Brondízio, H. T. Ngo, M. Guèze, J. Agard et al. Eds.). IPBES Secretariat, Bonn. https://ipbes.net/global-assessment

International Agency for Research on Cancer (IARC) and World Health Organisation (WHO). (2015). *IARC Monographs Volume 112: Evaluation of five organophosphate insecticides and herbicides*. https://www.iarc.who.int/wp-content/uploads/2018/07/MonographVolume112-1.pdf

International Programme on Chemical Safety. (1979). *DDT and its derivatives*. World Health Organisation. http://www.inchem.org/documents/ehc/ehc/ehc009.htm#SectionNumber:3.1

International Union for Conservation of Nature (IUCN). (2021). *The IUCN Red List of Threatened Species*. https://www.iucnredlist.org

Jacobson, M. F., & The Staff of the Center for Science in the Public Interest. (2006). *Six arguments for a greener diet: How a more plant-based diet could save your health and the environment.* Center for Science in the Public Interest.

Jasechko, S., & Perrone, D. (2021). Global groundwater wells at risk of running dry. *Science, 372*(6540), 418–421. https://doi.org/10.1126/science.abc2755

Kilvert, N. (2020). *Land clearing in Australia: How does your state (or territory) compare?* ABC News. https://www.abc.net.au/news/science/2020-10-08/deforestation-land-clearing-australia-state-by-state/12535438

Kissinger, G., Herold, M., & De Sy, V. (2012). *Drivers of deforestation and forest degradation: A synthesis report for REDD+ Policymakers*. Lexeme Consulting, Vancouver Canada. https://www.cifor.org/knowledge/publication/5167/

Less Meat Less Heat. (2017). *Facts*. http://www.lessmeatlessheat.org/facts/

Margat, J., & van der Gun, J. (2013). *Groundwater around the world: A geographic synopsis*. CRC Press/Balkema.

Marinova, D., & Raphaely, T. (2018). Impact of vegetarian diets on the environment. In W. Craig (Ed.), *Vegetarian nutrition and wellness* (pp. 13–24). CRC Press/Taylor and Francis.

Mekonnen, M. M., & Hoekstra, A. Y. (2012). A global assessment of the water footprint of farm animal products. *Ecosystems, 15*, 401–415. https://doi.org/10.1007/s10021-011-9517-8

Melhem, Y. B. (2021). *New Zealand's troubled waters*. ABC News. https://www.abc.net.au/news/2021-03-16/new-zealand-rivers-pollution-100-per-cent-pure/13236174

Millar, N. (2015). *Management of nitrogen fertilizer to reduce nitrous oxide emissions from field crops (E3152)*. Michigan State University, MSU Extension. https://www.canr.msu.edu/resources/management_of_nitrogen_fertilizer_to_reduce_nitrous_oxide_emissions_from_fi

Nation Master. (n.d.). *Nitrogen fertiliser production*. https://www.nationmaster.com/nmx/ranking/nitrogen-fertilizer-production

National Geographic. (2020). *Environmental impacts of agricultural modifications*. https://www.nationalgeographic.org/article/environmental-impacts-agricultural-modifications/

National Geographic. (n.d.). *A clean water crisis*. https://www.nationalgeographic.com/environment/article/freshwater-crisis?loggedin=true

National Pesticide Information Centre. (1999). *DDT (General Fact Sheet)*. http://npic.orst.edu/factsheets/ddtgen.pdf

National Snow & Ice Data Center (NSIDC). (2020). *Facts about glaciers*. https://nsidc.org/cryosphere/glaciers/quickfacts.html

Oppenlander, R. (2013). *Freshwater abuse and loss: Where is it all going?* https://www.forksoverknives.com/the-latest/freshwater-abuse-and-loss-where-is-it-all-going/

Owens, W., & Robinson, G. (2019). *Australia's deforestation rates at an all-time high*. The Junction. https://junctionjournalism.com/2019/09/20/australias-deforestation-rates-at-an-all-time-high/

Parikh, S. J., & James, B. R. (2012). Soil: The foundation of agriculture. *Nature Education Knowledge, 3*(10), 2. https://www.nature.com/scitable/knowledge/library/soil-the-foundation-of-agriculture-84224268/

Petel.bg. (2015). *We eat tomatoes sprayed with a banned poison*. https://petel.bg/YAdem-domati%2D%2Dpraskani-sas-zabranena-otrova__99811 [in Bulgarian].

Piesse, M. (2020). *Global water supply and demand trends point towards rising water insecurity*. Future Directions International. https://www.futuredirections.org.au/publication/global-water-supply-and-demand-trends-point-towards-rising-water-insecurity/

Pimentel, D., & Pimentel, M. (2003). Sustainability of meat-based and plant-based diets and the environment. *The American Journal of Clinical Nutrition, 78*(3), 660S–663S. https://doi.org/10.1093/ajcn/78.3.660S

Poore, J., & Nemecek, T. (2018). Reducing food's environmental impacts through producers and consumers. *Science, 360*(6392), 987–992. https://doi.org/10.1126/science.aaq0216

Potapov, P., Hansen, M. C., Laestadius, L., Turubanova, S., Yaroshenko, A., Thies, C., … Esipova, E. (2017). The last frontiers of wilderness: Tracking loss of intact forest landscapes from 2000 to 2013. *Science Advances, 3*(1), e1600821. https://doi.org/10.1126/sciadv.1600821

Ritchie, H. (2017). *How much of the world's land would we need in order to feed the global population with the average diet of a given country?* Our World in Data. https://ourworldindata.org/agricultural-land-by-global-diets

Ritchie, H., & Roser, M. (2019). *Land use*. Our World in Data. https://ourworldindata.org/land-use

Ritchie, H., & Roser, M. (2020). *Environmental impacts of food production*. Our World in Data. https://ourworldindata.org/environmental-impacts-of-food

Robbins, J. (2012). *How your food choices affect your health, happiness and the future of life on Earth* (25th anniversary edition). H J Kramer.

Rodell, M., Famiglietti, J. S., Wiese, D. N., Reager, J. T., Beaudoing, H. K., Landerer, F. W., & Lo, M.-H. (2018). Emerging trends in global freshwater availability. *Nature, 557*, 651–659. https://doi.org/10.1038/s41586-018-0123-1

Sánchez-Bayo, F., & Wyckhuys, K. A. G. (2019). Worldwide decline of the entomofauna: A review of its drivers. *Biological Conservation, 232*, 8–27. https://doi.org/10.1016/j.biocon.2019.01.020

Satterthwaite, D. (2020). *An urbanising world*. International Institute for Environment and Development. https://www.iied.org/urbanising-world

Secretariate of the Convention on Biological Diversity (CBD). (2020). *Global biodiversity outlook 5: Summary for policy makers*. https://www.cbd.int/gbo/gbo5/publication/gbo-5-spm-en.pdf

Sheals, J. G. (1953). Effects of DDT and BHC on soil arthropods. *Nature, 171*, 978. https://doi.org/10.1038/171978a0

Smith, P. (2012). Soils and climate change. *Current Opinion in Environmental Sustainability, 4*(5), 539–544. https://doi.org/10.1016/j.cosust.2012.06.005

Smith, P., Martino, D., Cai, Z., Gwary, D., Janzen, H., Kumar, P., … Smith, J. (2008). Greenhouse gas mitigation in agriculture. *Philosophical Transactions of the Royal Society of London. Series B, Biological Sciences, 363*(1492), 789–813. https://doi.org/10.1098/rstb.2007.2184

Spera, S. A., Winter, J. M., & Partridge, T. F. (2020). Brazilian maize yields negatively affected by climate after land clearing. *Nature Sustainability, 3*, 845–852. https://doi.org/10.1038/s41893-020-0560-3

Subbarao, G. V., Ito, O., Berry, W. L., & Wheeler, R. M. (2003). Sodium – A functional plant nutrient. *Critical Reviews in Plant Sciences, 22*, 391–416. https://doi.org/10.1080/07352680390243495

Terrer, C., Phillips, R. P., Hungate, B. A., Rosende, J., Pett-Ridge, J., Craig, M. E., … Jackson, R. B. (2021). A trade-off between plant and soil carbon storage under elevated CO_2. *Nature, 591*, 599–603. https://doi.org/10.1038/s41586-021-03306-8

The Macadamia. (2021). *Do we have enough water for all our macs?* https://themacadamia.co.za/2018/10/23/do-we-have-enough-water-for-all-our-macs/

Treu, H., Nordborg, M., Cederberg, C., Heuer, T., Claupein, E., Hoffmann, H., & Berndes, G. (2017). Carbon footprints and land use of conventional and organic diets in Germany. *Journal of Cleaner Production, 161*, 127–142. https://doi.org/10.1016/j.jclepro.2017.05.041

Troeh, F. R., & Thompson, L. M. (2005). *Soils and soil fertility* (6th ed.). Wiley-Blackwell.

United States Environmental Protection Agency (US EPA). (2020). *Understanding global warming potentials*. https://www.epa.gov/ghgemissions/understanding-global-warming-potentials

van den Berg, H., Manuweera, G., & Konradsen, F. (2017). Global trends in the production and use of DDT for control of malaria and other vector-borne diseases. *Malaria Journal, 16*, 401. https://doi.org/10.1186/s12936-017-2050-2

Vince, G. (2012). *Fertilisers: Enriching the world's soil*. BBC, Agriculture. https://www.bbc.com/future/article/20120828-enriching-the-soil

Von Alt, S. (2016). *Wake up! Tyson dumps over 6x more toxic pollution into waterways than Exxon*. Mercy for Animals. https://mercyforanimals.org/blog/wake-up-tyson-dumps-over-6x-more-toxic-pollution/

Wong, S. (2021). *Production of nitrogen fertiliser in China from 1990 to 2019*. Statista. https://www.statista.com/statistics/275858/nitrogen-fertilizer-production-for-chinas-agriculture/

World Bank. (2021). *Agricultural land (sq. km)*. https://data.worldbank.org/indicator/AG.LND.AGRI.K2

World Food Summit. (1996). *Food for all*. Rome. http://www.fao.org/3/x0262e/x0262e00.htm#TopOfPage

Worldometers. (2021). *Current world population*. https://www.worldometers.info/world-population/

Zomer, R. J., Bossio, D. A., Sommer, R., & Verchot, L. V. (2017). Global sequestration potential of increased organic carbon in cropland soils. *Scientific Reports, 7*, 15554. https://doi.org/10.1038/s41598-017-15794-8

Chapter 4
Reducing Food Waste and Packaging

Abstract The size of food wastage, including loss – during the production and supply chain, and waste – in retail and consumption, is a serious problem. Although food loss and waste differ across various parts of the world, the West is a major contributor to the problem. Food packaging adds to the severity of the situation as it heavily relies on plastics which contribute to microplastics pollution, which is spreading over the planet. The chapter covers the history of waste and discusses the need for reduction. Mimicking nature where there are no discarded materials is another strategy. Current developments to reduce waste and plastics-based packaging of food are also covered.

Estimates show that from all food calories produced in the world, approximately a quarter is wasted during the entire process from production, handling and storage, processing, distribution and sales to consumption (WRI, 2018). The size of food wastage, including loss – during production and supply chains, and waste – in retail and consumption, becomes one third if estimates are based on weight (WRI, 2018). This waste is not only reducing the opportunity for other people to have nutritional food but is also contributing to the heavy ecological footprint of food. Conservatively estimated, food wastage is responsible for 6% of total global greenhouse gas emissions, a share that is three times higher than emissions from aviation (Ritchie, 2020). A more accurate estimates of food waste alone (without food loss) show an even higher impact on climate change with 8–10% of global greenhouse gas emissions associated with foodstuffs that have been produced but not consumed (Mbow et al., 2019; UNEP, 2021).

Food loss and waste differ significantly across the world with wealthier countries throwing away excessively more at the consumption stage. For example, in 2011 only 5% of all food wastage happened in consumption in Sub-Saharan Africa compared to 61% in North America and Oceania where overall more than 40% of the food was lost or wasted (see Table 4.1). Latin America had the lowest food waste and loss of only 15%.

We both migrated to Australia during food shortages in Bulgaria at the time when the socialist system was collapsing. Little did we know that we were moving

Table 4.1 Food loss and food waste by region, 2011

Region	Share of food loss and waste	Food loss and waste by stage					
		Production	Handling and storage	Processing	Distribution and market	Consumption	Total
North America and Oceania	42%	17%	6%	9%	7%	61%	100%
Europe	22%	23%	12%	5%	9%	52%	100%
Industrialised Asia	25%	17%	23%	2%	11%	46%	100%
North Africa, West and Central Asia	19%	23%	21%	4%	18%	34%	100%
Latin America	15%	28%	22%	6%	17%	28%	100%
South and Southeast Asia	17%	32%	37%	4%	15%	13%	100%
Sub-Saharan Africa	23%	39%	37%	7%	13%	5%	100%

Source of data: WRI (2018)

to a society where food was uncritically thrown away by people. Our Bulgarian mothers used to say: "Letting food spoil and throwing it away is blasphemy"; in Australia, this was normal. Although a lot of emphasis is put on losses during the production, storage and processing of food, there is wide social acceptance of wasting foodstuffs. To make matters worse, USA, Australia, New Zealand and Canada – the countries belonging to the North America and Oceania group with the highest shares of food loss and consumption waste, are also the ones with the highest intake of animal-based products, and meat in particular. The overuse of fertilisers to produce animal feed and human food, including the one being wasted, is impacting on the planetary boundaries. Whilst technology can help reduce losses up to the farm gate, so far we have not been able to find ways to control food waste by households. As the statistics in Table 4.1 show, this is a social challenge for the developed world. Food is being disposed of in its original packaging unopened, often when it is still good for consumption (e.g. within the expiry date) or just because we have changed our meal plans.

In the 1990s, a small book about how to deal with waste was published – *Waste and Want: A Social History of Trash* by Susan Strasser (1999). In her book, Strasser describes in fascinating details many applications of our waste, looking at almost everything from feeding kitchen waste to pigs to making carpets from fabric scraps. These home products actually provided an important flow of resources to early American factories (Strasser, 1992). Paper mills, for example, depended on the cotton waste that was crushed into pulp and used to make paper. Some of these practices continue nowadays through recycling shops and exports to low-income countries. For example, some bakeries supply charity shops with their products at the end of the day or initiatives, such as dumpster diving or organised food rescue groups (see Fig. 4.1), aim at reducing food waste and saving it from going into landfills where it contributes to greenhouse gas emissions. The scale of the domestic food waste in the West however is so big that these efforts cannot curb the magnitude of the environmental problems caused. In a planetary emergency where our soils, air and waters need to be protected, food waste by the rich is simply an irresponsible behaviour with dismal consequences for all.

To add to the severity of the problem, there is also the packaging of food. Overpackaging and use of plastics for the food to reach the consumer are contributing to the enormous microplastics pollution, spreading over oceans, lands, rivers and contaminating even precipitations. When we think of waste, we somehow associate it with an inferno, which is what the modern world creates with the garbage it produces every year. The move to a circular economy in the area of food, requires us to

Fig. 4.1 Food van delivering rescued food, Perth, Western Australia, 2021. (Photo: Authors)

cut the unnecessary waste and loss and transition to a circular agriculture, ecological packaging and better food choices. Stopping the ecologically destructive behaviour is a complex but feasible task which should be handled without the vagaries of political agendas and the promises for future technological solutions. It needs to be done now.

The reason why we are linking food waste and packaging currently dominated by plastics, is simply because in the developed world, the majority of discarded foodstuffs are at the stage of consumption, including by households. Food has to be presented in a convenient and appealing form to reach the consumers. The two issues become intertwined and creating a sustainable food future in a planetary emergency requires addressing both. We first briefly refer to the history of waste and discuss the need for reduction and to mimic nature in which there are no discarded materials. Then we analyse current developments to reduce waste and plastics-based packaging to transition to a waste-free future with zero food plastics.

History of Waste

People have been handling the waste generated by them for millennia. The first landfill development dated to 3000 BC, is found in Knossos on the island of Crete in Greece (Commercial Zone, 2020). Rubbish was dumped by the Bronze Age Minoan people in large holes dug into the ground and was covered with earth (Watson, 2014). In 2000 BC, simple methods of composting and recycling were developed in China. Ancient Athens adopted the first waste-related law 2500 years ago in 500 BC which required garbage to be dumped at least one mile (1.6 km) away from the city in a special municipality landfill site. The ancient Greeks aimed at preserving Athens' beauty and most importantly, prevent the spread of diseases and illness (Lindenlauf, 2000). During the Roman Empire, the first public service for collection and disposal of waste was established (Eniscuola Energy & Environment, n.d.). Since the start of these remarkable ancient approaches to garbage, humanity has created many more waste management systems and regulations, aimed not so much to preserve the natural environment, but to protect people's health.

Waste dumps are found around many ancient cities, which must have been smelly and sinister. People however looked after every single thing and piece of food trying to find the best use for it or a new purpose to make it reusable (Strasser, 1999). Food leftovers were made into soup, given to the poor or fed to the domestic animals. Because of thrift, people applied ingenuity and prudence to deal with remaining or unwanted materials. Almost all garbage that was created before the Industrial Revolution was biological and organic, originating from kitchens and workshops and included plant and animal remains as well as human waste. The Industrial Revolution marked the start of a big acceleration in the production of everything, including waste. Industrialised agriculture and food production also started to generate large amounts of wastage. Any management systems, laws and regulations that were put in place aimed at safeguarding people's well-being but at the same time allowed them to behave lavishly and wastefully.

Throughout history, there has always been wastage of food. Large quantities are lost during food storage or transportation from one place to another. The majority of food loss in Southeast Asia for instance, occurs as a result of issues with storage and transportation and there is relatively low waste during consumption (Houston, 2017). We are in the first time point in history where waste started to represent such a significant part of the entire food system. The developing world also caught up with waste generation at a distribution/market and consumer level with China and India now producing the largest amounts, respectively 92 and 69 million tonnes per year (see Table 4.2).

Different factors and behaviours determine food waste within the context of each country. Let's examine some of the nations with the highest per capita food waste (see Table 4.2). Nigeria has the highest food waste on a per capita basis, which is a consequence from underdevelopment, and is due mainly to the lack of cold storage for the African heat and poor road infrastructure for accessing markets, rather than households being wasteful (Tomson, 2018). A study of household food waste in Greece, the second highest in the world on a per capita basis and where only 2% of it is composted, concluded that at least 30% is easily avoidable (Abeliotis et al., 2015). This includes before meal preparation (e.g. vegetable peeling and meat trimmings), during cooking (e.g. spoiling of foodstuffs and excess ingredients) and after consumption (e.g. food prepared in excess, spoiled or no longer wanted). Moreover, a review of evidence suggests that when it comes to reducing the waste generated at the

Table 4.2 Household food waste by country, selected countries, 2020

Country	Total waste per year [million tonnes]	Per capita waste per year [kg]	Country	Total waste per year [million tonnes]	Per capita waste per year [kg]
Nigeria	38.96	189	Vietnam	7.40	76
Greece	1.48	142	Germany	6.28	75
Saudi Arabia	3.66	105	Pakistan	16.35	74
Australia	2.62	102	Colombia	3.56	70
Iraq	4.10	102	Italy	4.05	67
Kenya	5.32	99	Bangladesh	10.70	65
Mexico	12.12	94	China	92.12	64
Ethiopia	10.58	92	Japan	8.09	64
Malaysia	2.95	91	Brazil	12.75	60
France	5.55	85	USA	19.53	59
Canada	2.98	79	Poland	2.12	56
Indonesia	21.06	77	India	69.00	50
UK	5.23	77	South Africa	2.37	40
Spain	3.60	77	Russia	4.82	33

Source of data: Statista (2020) and Worldometer (2020)

household level, food wastage has the highest prevention potential (Cox et al., 2010). The Kingdom of Saudi Arabia has limited arable land and water resources, however food is heavily subsidised and taken for granted as it is abundantly available in grocery stores, restaurants, celebrations and other catering events (Baig et al., 2019). This results in high per capita wastage. Combined with very limited awareness about the environmental footprint of food, the prevailing attitude is that unused food needs to be discarded to make room for new and fresher supplies (Baig et al., 2019). In Australia, another country with high food waste (102 kg per capita), the government estimates that the total of food loss and waste reaches 300 kg per person per year contributing to significant economic and environmental impacts (DAWE, n.d.).

It is only in the twenty-first century that we started to witness measures encouraging people to reduce their household waste, including that generated from their eating habits. The "pay-as-you-throw" garbage collection systems are becoming increasingly used at municipal level (Commercial Zone, 2020). Since the 1980s, commercial composting started to be introduced which allows food waste (together with green waste from yards) to be collected and processed separately producing quality composts. The main drivers for introducing municipal level composting are landfill closures, environmental consciousness and cost savings (Altman, 2016). Although these schemes handle food waste better than landfill, they do not encourage people to reduce it and mainly decrease methane emissions and smell.

Food packaging adds to the problems with waste. Starting from humble beginnings in metal tins and glass bottles, food packaging now uses thousands of different materials heavily relying on plastics (Davies, 2016). Once considered a revolutionary idea to make a disposable packaging or containers to encourage convenience in purchasing and consumption, today we have realised that such "benefits" are pointlessly wasteful. The history of food waste shows that humanity has moved from a culture that treasured each edible item to a transition where foodstuffs are taken for granted. This is combined with a disposable culture, which see consumption as a patriotic act supporting the economy, status and well-being of the nation at the expense of other human beings and the planet. In wealthy countries, food waste is a manifestation of a consumerist society where everyone expects the benefits from waste management in dealing with unwanted items. This consumerist behaviour does not take into account the limits and implications for the planet and its natural resources, including the ability to continue to produce food. It is only in the early twenty-first century that food packaging started to be examined from an ecological point of view in the search of innovative solutions.

Reduction Targets

The road from the farm to the fork is long and during this process, one third of the world's food or 1.3 billion tonnes of food yearly that could feed about 3 billion people do not reach people's plates at all (FAO, 2019; UNEP, n.d.). The most concerning fact is that this is almost 4 times the number of all starving people in the world. Moreover, there are huge resources associated with this wastage. During the production cycle, 25% of the water used in agriculture to grow food is also ultimately wasted, because of non-utilisation of the final harvest (DAWE, n.d.). As already stated, food waste produces 8–10% of the global greenhouse gas emissions and if it were a country, it would be the third largest emitter, behind China and USA (DAWE, n.d.).

Food wastage amounts to US\$ 310 billion annually in developing countries, the bulk of it being from fruits and vegetables (UNEP, n.d.). In the developed world, food losses and waste are more than double that amount at US\$680 billion (UNDP, n.d.). Developed and developing countries dissipate roughly the same amounts of food, respectively 670 and 630 million tonnes annually (UNEP, n.d.). The share of animal-based discarded foods in the industrialised countries, however, is much higher. In USA, for example, meat, poultry, fish and dairy account for 55% of the country's food waste at the level of retailer and consumer while fruit and vegetables represent 26% (Buzby & Hyman, 2012). The food loss and food waste in the meat sector in the European Union are estimated to be 23% (Karwowska et al., 2021). Globally wastage of meat and dairy is at around 20% each (FAO, 2015), but it accounts for 60% of the land footprint associated with food loss and waste (FAO, 2019).

Some will say that after all, food waste is organic and therefore, it should not be such a big environmental issue. It is true that food is naturally biodegradable, but there are many caveats. Firstly, discarded food products need to reach the right environment as if disposed or mixed with other inorganic wastes, they are no longer biodegradable. Secondly, packaging which contains plastics significantly alters the biodegradability of food. Thirdly, there have already been greenhouse gas emissions associated with the production of food and there will be more as a result of its decomposition. Finally, and most importantly, a lot of precious and limited resources, many of them non-renewable, have already been used to produce the food that ultimately is not being consumed.

The majority of studies and policy documents set the target of halving food waste and loss by 2030. This is one of the targets (Target 12.3) of Sustainable Development Goal 12 Responsible Consumption and Production (FAO, 2021). Similar calls are being made by Green Peace (Greenpeace, 2018; Tirado et al., 2018), the EAT-Lancet Commission on Planetary Health (Willett et al., 2019) and the United Nations Food Systems Summit (United Nations, 2021). The US also has the same target (FoodPrint, 2021). These are not very ambitious aims given the existing evidence that food waste is relatively easy to prevent. Possible avenues include improved awareness, preventing overbuying, reducing human errors in judgement and planning, better technological solutions related to retail efficiencies and logistics.

A consideration that should also be included in such targets is that ultimately the consumers are the ones who set the rules. The halving of food waste can easily be sped up if the consumers change their attitudes as to what is acceptable. Nutritional considerations should prevail above aesthetic perceptions as to what makes the perfect carrot or banana, particularly for those living in wealthier countries.

Mimicking Nature

Waste, including food waste and packaging, is by all means of anthropogenic nature. If you look deeply into this, you could discover many connotations giving "waste" a meaning. From unwanted and not used to useless, careless, extravagant and indulgent consumption, food waste is a bigger problem than dealing with satisfying human hunger. The word also means dissipation, destruction, atrophy and death (Strasser, 1992). These are all implications of waste. However, there is no waste in nature. No materials are wasted in the natural environment as they get reused through feedback cycles. Things in nature are biodegradable, they decompose, everything made by nature returns to nature. Natural systems operate in a constantly rotating, closed cycle and people should be part of this.

In reality however, we have created and use a lot of chemicals and plastics in the ways we grow and package food, which is causing environmental deterioration as many of these products cannot be absorbed in nature's natural cycles. If we are to reduce and eliminate food waste, we also need to mimic the way nature operates. Everything it creates has a well-designed packaging, for example, cucumbers or apples are protected by their skins from germs and various environmental influences. Supermarkets try to extend the shelf life of fruits and vegetables by using plastics and although this may produce some positive results in reducing waste, we should be looking at packaging solutions with which nature can cope through its biological cycles.

Biomimicry and ecomimicry have emerged as ways that we humans should think about handling food and use regenerative solutions that optimise resource use and most importantly, return nutrients back to the soil. Rather than thinking what we can take from nature, we need to switch to how we can learn from the natural processes and cycles (Biomimicry Institute, 2021). We already have some solutions, such as the use of cyanobacteria to create closed-loop hydroponics, sustainable edible packaging, collecting atmospheric water to irrigate plants or food preservation methods without the use of electricity from the main grid, natural encapsulation of plants' natural defence molecules, such as volatile organic compounds, to fight against fungi, use of essential oils, such as made from oregano, eucalyptus or lavender, to protect against pathogens, apps that allow people to make optimal food purchases... (Biomimicry Institute, 2021). For example, one of the ecomimicry innovations developed by the Pratt Institute (2019) is to prevent the spoilage of tomatoes in Nigeria, the biggest producer in Sub-Saharan Africa, and uses locally available eco-friendly materials. Evaporative colling which imitates the body's reaction to heat through perspiration is also put forward as a solution (Nair & Lim, 2017).

Although much deeper changes are needed to transform the current food systems, the important thing is to change our way of thinking and explore opportunities inspired by the biophysical world. It is also extremely important to understand how nature deals with pollution, contamination, pests or spoilage and develop biological strategies. Imitating the way the natural world works through biomimicry and ecomimicry should be the best design approach in our efforts to reduce and avoid food loss and waste as well as to deal with the issue of packaging.

Zero Food Plastics

Back in 1907, humanity was excited and amazed by the news about Leo Baekeland's invention of Bakelite, the first modern synthetic polymer and predecessor of today's multitude of plastics. They were predicted to be the "material of 1000 uses" (García, 2016; Rhodes, 2018). New inventions followed with polyvinyl chloride (known as PVC or vinyl), polyethylene (particularly suited for shopping bags, food containers and drink bottles) and polypropylene (similarly widely used for packaging). Plastics were indeed a great novelty and there was a lot of hope and many newly created opportunities. After the end of the Second World War, the food industry adopted plastics on a large scale as these products were versatile, easy to work with and perceived as clean. Together with the food industry, all other economic sectors were encouraged to embrace a throwaway culture to stimulate the economy. This generated profits for the petrochemical companies while billions of plastic items were disposed in landfills, incinerated or just dumped. The production of plastics has been steadily on the increase from 2 million tonnes in 1950 to 368 million tonnes in 2019 (PlasticsEurope, 2020).

Currently packaging of drinks and food is responsible for the lion share of plastics use with a lot of waste generated by a relatively low number of companies. For example, in 2019 Coca-Cola produced 3 million tonnes of packaging waste, including 88 billion plastic bottles (or 167,000 per minute) which if lined up end to end would reach to the Moon and back 31 times (Plastic Atlas, 2019). Other major plastics polluters are Nestlé, Danon and Unilever. Plastic films and foams make food look more attractive and protect it from damage while keeping it fresh (Plastic Atlas, 2019). Although plastics are recyclable, in reality the majority ends in landfill or is incinerated – a process described as energy recovery which destroys materials with a lot of embedded energy to generate a small amount of power. In Europe, for example, 25% of all post-consumer plastic waste ends up in landfill (PalsticsEurope, 2020).

The convenience of plastics started to fade in comparison with all known and unknown problems they have generated for the natural environment and human health. Examples include bisphenol A (BRA) associated with toxicological effects on brain health, potential carcinogen (Gao et al., 2015; Khan et al., 2021) and causing immune and reproduction system damage (Huo et al., 2015). The irresponsible discarding of plastics has resulted in sea animals being dangerously wrapped in this material (Rhodes, 2018) and 14 million tonnes of microplastics accumulated on the seafloor (Lau, 2021), increasing the chances of dying for 22% of turtles, affecting 95% of all seabirds and 600 species overall, from microorganisms to whales (Secretariat of the CBD, 2012; Wilcox et al., 2015; Lau, 2021). The plastic manufacturers themselves admit that: "Plastic waste is unacceptable in any habitat" (PlasticsEurope, 2020, p. 5).

However, instead of dealing with the problem, the majority of the western countries have been exporting the plastics to developing countries, including plastic bottles, packaging and food containers. In 2018, China stopped accepting foreign recycling waste as part of its National Sword policy, no longer wanting to act as a dumping ground processing half of the world's recyclable plastics (Katz, 2019) and imposing the need for the exporting countries to find ways to deal with their own waste. Even before the ban, China was recycling only 9% of the discarded plastics, another slightly bigger amount was burned and the rest was buried in landfills or simply dumped and left to wash away in rivers and oceans (Brooks et al., 2018; Katz, 2019). China was not the only country made to deal with the world's unwanted plastic waste, including that generated by the food industry. The higher-income countries from the Organisation for Economic Cooperation (OECD) have been exporting their plastic waste (70% of it in 2016) to lower-income countries in East Asia and the Pacific for decades (Brooks et al., 2018).

The price of the temporary convenience and the ease of passing on the problem to others, resulted in a serious planetary health problem. Enormous quantities of non-reusable and not recycled plastic stuff severely contaminated the global environment creating an ecological catastrophe (Jambeck et al., 2015). The scale of the problem is immense given that 56% of all plastics in the world have been made between 2000 and 2020 and the expectation is that production will further increase to 600 million tonnes by 2025 (Plastic Soup Foundation, n.d.). On land and in the marine environment, plastics also break down in size forming debris of microplastics – microbeads usually less than 5 mm long, which are mistakenly consumed by animals. Plastics are no longer only part of packaging, they have now entered the food chain. Through trophic transfer – animals eating other species who have ingested microplastics, people are also consuming them. In fact, the risk of bioaccumulation of toxins through exposure to the plastics themselves is much smaller compared to that through the food chain.

While there are some areas where the use of plastics may be unavoidable, better solutions should be applied in the food industry. In all areas national policies encouraging recycling and whenever possible, banning plastics and the export of plastics waste for processing abroad (as is the case in UK, Sharp, 2020) are urgently needed. In addition to food packaging, agriculture is a major user of plastics – estimated at 6.5 million tonnes of the material globally each year (Plastic Atlas, 2019). Without a policy change, the mountains of rubbish comprising 8 billion tonnes of indestructible plastics produced somewhere worldwide in the last six decades will continue to grow (Katz, 2019) blocking sewages, releasing toxic chemicals, creating breading grounds for diseases and killing animals and other species. Although many companies are committing to reducing the use of plastics and replacing them with biopolymers (bioplastics), we must not delude ourselves that this will happen quickly and will result in immediate changes. Global plastic production is still projected to increase by 36% or more over the next 5 years and potentially extend this increase until 2030 (Lau, 2021). Recycling becomes an essential avenue of containing the spread of plastics within a circular economy. This is even more important as a result from the COVID-19 pandemic during which the use of disposable gloves, protective clothes, face masks, extra syringes for vaccinations etc. which additionally contribute to the large amount of already accumulated plastic waste worldwide.

Zero food plastics may seem like an impossible task. However, the first step is to reduce or eliminate food waste which will cut down significantly the need to use plastic. The second is to replace single-use containers with more durable ones. However, the most valuable approach is to eliminate the use of synthetic plastics that do not naturally disintegrate. In fact, the popular 3R (Reduce, Reuse, Recycle) hierarchy of recycling should be replaced with 5Rs – Refuse, Reduce, Reuse, Recycle, and Rot (Johnson, 2013). By refusing the temptation to purchase or possess unnecessary food or packaged items, such as disposable coffee cups, single use plastic containers, utensils, cutlery and straws, the task of dealing with waste becomes easier. Reducing allows you to give thought of what you really need and want while also considering the way it is packaged. Natural materials, such as paper or bamboo, are an environmentally better option. Reuse allows you to extend the life of any packaging, containers and even food leftovers which can be used for a different meal. Recycling within this context applies not only to any food-related plastics but also to composting the organic materials, such as kitchen scrap. For the final stage – rot, you do not need to do a lot if you have composting services organised by your local council or even better, if you have your own composting bin, bay or pile depending on the availability of space. The worms will do the work for you producing what some call "black gold" – compost which is an excellent soil conditioner that adds nutrients and helps with moisture retention (Good Living, 2021). It is essential to avoid any plastics as well as materials that have been chemically treated.

Currently there is little economic value in recovering the plastics from the environment (Lau, 2021). However recycling and reuse of plastics is essential in many of the new ways of growing food, such as vertical farming and various ponics methods. It is expected innovative technologies to be able to help clean and prevent further pollution of the terrestrial, riverine and marine environments using artificial intelligence, machine learning and other smart, especially designed tools, such as floating rubbish bins (Lau, 2021). Community members and watchdog organisations, such as Sea Shepard or the US Operation Clean Sweep, play a vital role in monitoring where plastic waste goes. Edible packaging is another area where possible progress is expected along the lines of biomimicry (Mouradian, 2017). Social marketing can also play an important role in people developing the right attitudes towards reducing food packaging (see Fig. 4.2).

Fig. 4.2 Recycling collection track advertising reduction in food packaging, Perth, Australia, 2021. (Photo: Authors)

Waste-Free Food Future

Creating a world without waste is an ambitious plan, especially related to plastics, but not impossible when it comes to food. Agricultural produce can be recovered, rescued, repurposed, recreated and even sold on existing markets, meaning that with more thoughtful strategies and creating synergies, we can make a dramatic shift toward reducing food waste and even achieving zero food waste.

In Australia, for instance, food rescue organisations, such as OzHarvest (see Fig. 4.1) and Foodbank, are dedicating their efforts to educating the community and preventing food waste practices, collecting food from supermarkets and transporting it to places where they can donate it to people in need. Fruits and vegetables, often discarded for aesthetic reasons, can be salvaged and made into fresh juice for vending machines around the big cities. The COVID-19 pandemic brought a big shock to the world and the food systems, often diminishing the opportunities to eliminate food waste, particularly when restaurants, coffee shops and other catering places were closed at a very short notice. However, this health emergency also delivered innovative models of food and meal kits delivery directly to people's homes, particularly of seasonal produce. Examples in Australia are HelloFresh, Marley Spoon and Dinnerly as well as supplies directly from organic farms. Such initiatives also encourage people to reconnect with cooking at home and appreciate the true value of food where nothing is wasted.

Concluding Comments

A 1576 English proverb states: "Wilful waste brings woeful want" (Speake, 2015). Are we heading towards such a food situation in the future? We have many reasons to believe that this is the case, unless we change our behaviour. In order to deal with waste, we need to also understand want (Strasser, 1992). We need to change the nature of food consumption, its structure and strictures. The COVID-19 pandemic allowed us to develop some innovative ways of eliminating food waste but a lot more work needs to be done.

While some national contributions to the Paris Agreement mention food loss, none talks about food waste (Schulte et al., 2020) despite the calls for a global movement to transform our food systems making them safe, accessible, sustainable and equitable. Eliminating food waste in the wealthy countries is within the reach of most of their citizens. It only requires good will to deal with the planetary emergency and avoid woeful global want.

References

Abeliotis, K., Lasaridi, K., Costarelli, V., & Chroni, C. (2015). The implications of food waste generation on climate change: The case of Greece. *Sustainable Production and Consumption, 3*, 81–84. https://doi.org/10.1016/j.spc.2015.06.006

Altman, V. (2016). *Municipal composting schemes: International case studies.* Cooperative Research Centre Low Carbon Living. https://apo.org.au/sites/default/files/resource-files/2016-11/apo-nid240311.pdf

Baig, M. B., Al-Zahrani, K. H., Schneider, F., Straquadine, G. S., & Mourad, M. (2019). Food waste posing a serious threat to sustainability in the Kingdom of Saudi Arabia – A systematic review. *Saudi Journal of Biological Sciences, 26*(7), 1743–1752. https://doi.org/10.1016/j.sjbs.2018.06.004

Biomimicry Institute. (2021). *Six revolutionary ideas to change the way we feed the planet, inspired by nature.* https://biomimicry.org/six-revolutionary-ideas-feed-the-planet/

Brooks, A. L., Wang, S., & Jambeck, J. R. (2018). The Chinese import ban and its impact on global plastic waste trade. *Science Advances, 4*(6), eaat0131. https://doi.org/10.1126/sciadv.aat0131

Buzby, J. C., & Hyman, J. (2012). Total and per capita value of food loss in the United States. *Food Policy, 37*(5), 561–570. https://doi.org/10.1016/j.foodpol.2012.06.002

Commercial Zone. (2020). *A brief history of waste management.* https://www.commercialzone.com/a-brief-history-of-waste-management/

Cox, J., Glorgi, S., Sharp, V., Strange, K., Wilson, D. C., & Blakey, N. (2010). Household waste prevention – A review of evidence. *Waste Management & Research, 28*(3), 193–219. https://doi.org/10.1177/0734242X10361506

Davies, M. (2016). *The history of food packaging.* Charlotte Packaging. https://www.charlotte-packaging.com/latest-news/history-food-packaging/

Department of Agriculture, Water and the Environment (DAWE). (n.d.). *Tackling Australia's food waste*. Australian Government. https://www.environment.gov.au/protection/waste/food-waste#:~:text=Each%20year%20we%20waste%20around,of%20Australia's%20greenhouse%20gas%20emissions

Eniscuola Energy & Environment. (n.d.). *Waste in pre-industrial society...* http://www.eniscuola.net/en/argomento/waste/what-is-waste/waste-in-pre-industrial-society/

Food and Agriculture Organisation of the United Nations (FAO). (2015). *Food loss and waste*. http://www.fao.org/3/i4807e/i4807e.pdf

Food and Agriculture Organisation of the United Nations (FAO). (2019). *The state of food and agriculture 2019: Moving forward on food loss and waste reduction*. Rome. http://www.fao.org/3/ca6030en/ca6030en.pdf

Food and Agriculture Organisation of the United Nations (FAO). (2021). *Sustainable Development Goals*. http://www.fao.org/sustainable-development-goals/indicators/1231/en/

FoodPrint. (2021). *The problem of food waste: Almost half of our food is wasted in the United States. How does this happen? What can we do to solve our enormous food waste problem?* https://foodprint.org/issues/the-problem-of-food-waste/

Gao, H., Yang, B. J., Li, N., Feng, L. M., Shi, X. Y., Zhao, W. H., & Liu, S. J. (2015). Bisphenol A and hormone-associated cancers: Current progress and perspectives. *Medicine, 94*(1), e211. https://doi.org/10.1097/MD.0000000000000211

García, J. M. (2016). Catalyst: Design challenges for the future of plastics recycling. *Chem, 1*(6), 813–815. https://doi.org/10.1016/j.chempr.2016.11.003

Good Living. (2021). *A beginner's guide to composting*. Government of South Australia. https://www.environment.sa.gov.au/goodliving/posts/2019/05/guide-to-composting

Greenpeace. (2018). *Less is more: Reducing meat and dairy for a healthier life and planet. The Greenpeace vision of the meat and dairy system towards 2050*. https://www.greenpeace.org/static/planet4-international-stateless/2018/03/698c4c4a-summary_greenpeace-livestock-vision-towards-2050.pdf

Houston, G. (2017). *This ancient storage technique could be the solution to food waste*. Food and Wine. https://www.foodandwine.com/news/ancient-storage-technique-could-be-solution-asias-food-waste-issue

Huo, X., Chen, D., He, Y., Zhu, W., Zhou, W., & Zhang, J. (2015). Bisphenol-A and female infertility: A possible role of gene-environment interactions. *International Journal of Environmental Research and Public Health, 12*, 11101–11116. https://doi.org/10.3390/ijerph120911101

Jambeck, J. R., Andrady, A., Geyer, R., Narayan, R., Perryman, M., Siegler, T., Wilcox, C., & Law, K. L. (2015). Plastic waste inputs from land into the ocean. *Science, 347*(6223), 768–771. https://doi.org/10.1126/science.1260352

Johnson, B. (2013). *Zero waste home: The ultimate guide to simplifying your life by reducing your waste*. Scribner.

Karwowska, M., Łaba, S., & Szczepański, K. (2021). Food loss and waste in meat sector – Why the consumption stage generates the most losses? *Sustainability, 13*, 6227. https://doi.org/10.3390/su13116227

Katz, C. (2019). *The world's recycling is in chaos. Here's what has to happen*. Wired. https://www.wired.com/story/the-worlds-recycling-is-in-chaos-heres-what-has-to-happen/

Khan, N. G., Correia, J., Adiga, D., Satwadi Rai, P., Souza, H. S., Chakrabarty, S., & Kabekkodu, S. P. (2021). A comprehensive review on the carcinogenic potential of bisphenol A: Clues and evidence. *Environmental Science and Pollution Research, 28*, 19643–19663. https://doi.org/10.1007/s11356-021-13071-w

Lau, D. (2021). *Reimagining the future of plastics*. ECOS, CSIRO. https://ecos.csiro.au/reimagining-the-future-of-plastics/

Lindenlauf, A. (2000). *Waste management in ancient Greece from the Homeric to the Classical period: Concepts and practices of waste, dirt, recycling and disposal*. Doctoral thesis, University of London. https://discovery.ucl.ac.uk/id/eprint/1317693

Mbow, C., Rosenzweig, C., Barioni, L. G., Benton, T. G., Herrero, M., Krishnapillai, M., … Xu, Y. (2019). Chapter 5. Food security. In *Climate change and land: An IPCC special report on climate change, desertification, land degradation, sustainable land management, food security, and greenhouse gas fluxes in terrestrial ecosystems* (pp. 437–550). Intergovernmental Panel on Climate Change. https://www.ipcc.ch/site/assets/uploads/sites/4/2021/02/08_Chapter-5_3.pdf

Mouradian, N. (2017). *11 examples of packaging you can actually eat*. Dieline. https://thedieline.com/blog/2017/10/16/11-examples-of-packaging-you-can-actually-eat?

Nair, T., & Lim, C. (2017). *Science, technology and human security – Fighting food wastage: New ideas from the past*. RSIS Commentaries No. 016. Nanyang Technological University. https://dr.ntu.edu.sg/bitstream/10220/42051/1/CO17016.pdf

Plastic Atlas. (2019). *Facts and figures about the world of synthetic polymers* (2nd ed.). Heinrich Böll Foundation, Berlin, Germany & Break Free From Plastic. https://www.boell.de/sites/default/files/2020-01/Plastic%20Atlas%202019%202nd%20Edition.pdf?dimension1=ds_plastikatlas

Plastic Soup Foundation. (n.d.). *Facts & figures*. https://www.plasticsoupfoundation.org/en/plastic-facts-and-figures/

Plastics Europe. (2020). *Plastics – The facts 2020: An analysis of European plastics production, demand and waste data*. https://www.plasticseurope.org/en/resources/publications/4312-plastics-facts-2020

Pratt Institute. (2019). *Adaptations in nature inspire students in award-winning design to reduce food waste*. https://news.pratt.edu/article/adaptations-in-nature-inspire-students-in-award-winning-design-to-reduce-fo/

Rhodes, C. J. (2018). Plastic pollution and potential solutions. *Science Progress, 101*(3), 207–260. https://doi.org/10.3184/003685018X15294876706211

Ritchie, H. (2020). *Food waste is responsible for 6% of global greenhouse gas emissions*. Our world in data. https://ourworldindata.org/food-waste-emissionss

Schulte, I., Bakhtary, H., Siantidis, S., Haupt, F., Fleckenstein, M., & O'Connor, C. (2020). *Enhancing NDCs for food systems: Recommendations for decision-makers*. WWF Germany & WWF Food Practice. https://www.climatefocus.com/sites/default/files/200909_WWF_NDC_Food_final_low.pdf

Secretariat of the Convention on Biological Diversity (CBD). (2012). *Impacts of marine debris on biodiversity: Current status and potential solutions*. Technical Series No. 67. Montreal. https://www.cbd.int/doc/publications/cbd-ts-67-en.pdf

Sharp, A. (2020). *Can we stop offshoring our plastic problem?* Phys.org. https://phys.org/news/2020-02-offshoring-plastic-problem.html?utm_source=TrendMD&utm_medium=cpc&utm_campaign=Phys.org_TrendMD_1

Speake, J. (Ed.). (2015). *Oxford dictionary of proverbs* (6th ed.). WILFUL waste makes woeful want. https://www.oxfordreference.com/view/10.1093/acref/9780198734901.001.0001/acref-9780198734901-e-2456

Statista. (2020). *Annual per capita household food waste of selected countries worldwide as of 2020 (in kilograms per year)*. https://www.statista.com/statistics/933059/per-capita-food-waste-of-selected-countries/

Strasser, S. (1992). *Waste and want: The other side of consumption*. Berg Publishers. https://www.ghi-dc.org/fileadmin/publications/Annual_Lecture_Series/Waste_and_Want.pdf

Strasser, S. (1999). *Waste and want: A social history of trash*. Henry Holt & Co.

Tirado, R., Thompson, K. F., Miller, K. A., & Johnston, P. (2018). *Less is more: Reducing meat and dairy for a healthier life and planet*. Greenpeace Research Laboratories Technical Report (Review) 03–2018. https://www.greenpeace.org/international/publication/15093/less-is-more/

Tomson, B. (2018). *Food waste in Africa starts long before the grocery store*. Agri-Pulse. https://www.agri-pulse.com/articles/11672-food-waste-in-africa-starts-long-before-the-grocery-store

United Nations. (2021). *The 2021 Food Systems Summit*. https://www.un.org/en/food-systems-summit

United Nations Environment Programme (UNEP). (2021). *Food Waste Index Report 2021*. Nairobi. https://www.unep.org/resources/report/unep-food-waste-index-report-2021?fbclid=IwAR37F YkMsgNj9Fb68ONjVW3A6IOXC1CuDrluR2rPN6cfw-3uby9ZGKH7c9Q

United Nations Environment Programme (UNEP). (n.d.). *Worldwide food waste*. https://www. unep.org/thinkeatsave/get-informed/worldwide-food-waste

Watson, J. (2014). *Recycling in the age of Hercules*. Halton Recycles. https://haltonrecycles.word-press.com/2014/01/09/recycling-in-the-age-of-hercules/

Wilcox, C., Sebille, E. W., & Hardesty, B. D. (2015). Threat of plastic pollution to seabirds is global, pervasive, and increasing. *Proceedings of the National Academy of Sciences of the United States of America (PNAS), 112*(38), 11899–11904. https://doi.org/10.1073/pnas.1502108112

Willett, W., Rockström, J., Loken, B., Springmann, M., Lang, T., Vermeulen, S., … Murray, C. J. L. (2019). Food in the Anthropocene: The EAT–Lancet Commission on healthy diets from sustainable food systems. *The Lancet, 393*(10170), 447–492. https://doi.org/10.1016/S0140-6736(18)31788-4

World Resources Institute (WRI). (2018). *Creating a sustainable food future: A menu of solutions to feed nearly 10 billion people by 2050*. Synthesis Report. World Bank, UN Environment, UNDP, CIRAD, INRA.

Worldometer. (2020). *Current world population*. https://www.worldometers.info/world-population

Part II
Industry and Marketing Perspectives

Chapter 5
Circular Agriculture

Abstract Circular agriculture is defined. Five underlying principles, namely regeneration, land-effectiveness, integrated management, focus on quality food and localisation, are outlined together with comments on what is necessary for a transition to circularity from the current linear agricultural model. The overall message of the chapter is that it is possible to provide quality food while recycling nutrients within agriculture to feed not only humans but also the soil, its microbes and other living species on this planet. Circular agriculture offers most-needed solutions to dealing with the planetary emergency.

We are writing this book from Curtin University's Legacy Living Lab (Curtin University, n.d.). This is a building that prides itself for demonstrating the principles of a circular economy and reducing the greenhouse gas emissions associated with construction and use. Designed from modules which can be disassembled and reused past the current three-year projected life of the Lab in Fremantle, Western Australia, the building offers a comfortable high-tech working environment. These days almost everybody has heard about the concept of a circular economy which replaces the previous linear model of "take–make–use–dispose" with "make–use–return" (Zaman & Ahsan, 2019). The building of the Lab is a step in the right direction and demonstration that things can be done better while still providing a quality office space for its users. With emphasis on energy, construction materials, water consumption and an electric car parked outside for shared use, the Lab is an example of urban sustainability research. It draws together important elements of sustainability but excludes the basic need for human survival – food.

Ironically, nowhere has the circular economy concept manifested better than in the field of agriculture. In fact, traditionally before the invent of artificial fertilisers and the globalised system of food production, the only practised agricultural model was circular. Farmers grew plants for food and feed, raised farm animals and after the process of consumption, the residues were returned back to the land. Circularity was similarly represented through the cycle of seasons associated with different agricultural activities. With the advent of modern agriculture, this circular model

D. Marinova, D. Bogueva, *Food in a Planetary Emergency*,
https://doi.org/10.1007/978-981-16-7707-6_5

was disrupted to the point of becoming unrecognisable. Through the application of modern technologies, humanity has been able to drastically improve the quantity of the food produced. This however has come at the expense of the quality of the food in terms of its nutritional values and taste, but more alarmingly it had contributed towards reduced fertility of the soils available on this planet (Smith et al., 2016; Timmis & Ramos, 2021). According to the Ellen MacArthur Foundation (2017), for every $1 spent on food, there are $2 environmental, health and economic costs to society. Economists describe these costs as negative externalities which are not captured by the market mechanisms we currently use. Although on the surface it may appear that the benefits from these negative externalities go to those who produce and consume the food, the reality is that people rarely think about the price we pay to satisfy human hunger.

Urbanisation has drawn people away from rural areas where the majority of food is produced resulting in a disconnect between what we eat and where it comes from. In many ways, the food production system has become invisible and hidden from the urban dweller who mainly sees the final product as a consumer in the shops. Although the popularity of farmers' markets and community gardens has increased in the twenty-first century, many questions surrounding agricultural systems remain veiled and concealed from those who live in the cities expected to reach 85% of the global population by 2100 (OECD, 2015). In Australia, where the circular economy demonstration lab is, the share of urban population as of 2020 is already 86% (World Bank, 2021). The Lab has also excluded food from its circular economy model. Is it because the production and consumption of foodstuff are a necessary evil that is too difficult to handle or is it because we do not care about food? The answer is: none of the above. Circular agriculture is a way of producing food that should be understood on its own merit. Although it can play a role within a circular economy, food production is an essential process for human existence and should be able to function on it is own without overexploiting the resources of this planet.

Many studies and modelling (Willett et al., 2019; Crippa et al., 2021) show that the modern system of food production requires drastic changes in order to satisfy the double priorities of feeding the global population while preserving the health of the environment. Back in 2010, a Canadian study by Pelletier and Tyedmers (2010) showed that irrespective of the progress we make in the other sectors of the economy, including energy, construction, buildings and industry, food production single-handedly will make life on this planet unbearable if not completely changed. The linear model of production of food (as well as any other items for human consumption) is equivalent to exploitation of the natural environment and taking it for granted until it can no longer deliver the ecological functions we need for our survival as a species. To respond to the alarming call for a planetary emergency, a return to a circular agricultural model is seen as a solution. Is this still possible and how?

In this chapter, we first present a brief definition of circular agriculture, then we elaborate on the underlying principles and comment on what is necessary for a transition to circularity from the current linear model. The issues surround waste and food production are also discussed. We conclude by commenting on the implementation of circular agriculture. The overall message is that it is possible to provide quality food while recycling nutrients within agriculture to feed not only humans but also the soil, its microbes and other living species that co-exist with us on this planet.

Defining Circular Agriculture

A technical description of circular agriculture is a closed-loop system in which nothing is wasted (Toop et al., 2017). Another explanation of circular agriculture is a system that mimics the natural processes of regeneration (Ellen MacArthur Foundation, 2017). A most important aspect of such definitions is related to the boundaries of the system. In the strictest sense, the food waste or by-products should be returned to the soil as organic fertilisers to maintain its productivity. The majority of approaches however keep these boundaries quite loose and in addition to animal feed and biofertilisers, include the production of biofuels, biochemicals, biopolymers, nutraceuticals and pharmaceuticals whose applications are outside the food sector. Some even argue that the longer the industrial chain of application of agriculture-based products, the better for the environment. This however is not supported by empirical evidence and different combinations of agriculture-based products and applications can deliver optimal environmental performance depending on the circumstances (Fan et al., 2018).

The way the question about system boundaries can be resolved is to distinguish between a circular economy in which agriculture plays a significant role and circular agriculture where the closed-loop cycle relates only to food production. It is the latter approach that is of interest here because of our specific focus on food. Such an approach also allows us to justify the fact that the circular economy Legacy Living Lab does not incorporate food. In other words, creating circular models tailored at different sectors of the economy can be a promising way to achieving sustainability and responding to the planetary emergency.

Also, circular agriculture would look very differently when applied in Australia compared to Europe, Asia or any other place. In fact, some argue that universal solutions are hard to develop and each farmer should find what works for the specific circumstances of their farm (Schouten, 2020). Before defining what circular agriculture means, the Wageningen University in the Netherlands specifies what it is not – it is not a return to pre-industrial ways of producing food or imposition of strict government regulations and market requirements to farmers (Wageningen University & Research, 2018). It is about using the residuals of agricultural biomass and from food processing as renewable resources that maintain soil productivity. The health of soils is a major aim in circular agriculture and this could be achieved through the cycling of organic matter which provides nutrients, such as phosphorus, nitrogen, potassium, trace elements and many micronutrients.

According to Muscio and Sisto (2020), circularity in agriculture also means minimal use of external inputs for the production of agricultural commodities. This implies that agricultural systems should be maintained through closed nutrient loops which cover crops and livestock and minimise negative discharges in the environment, including water bodies and the atmosphere. Through circular agriculture, food production needs to be "decoupled" from environmental deterioration (Esposito et al., 2020).

Transitioning to and maintaining a circular agriculture way of food production are a process that requires a lot of innovation informed by research. The final aim is to achieve a global shift through localised solutions and local closed-loop systems. However, certain common principles should be followed in this process.

Principles of Circular Agriculture

Below are five principles which provide a shared ground for circular agriculture. Some of them may appear as common sense, but the changes the world has witnessed since industrialisation and intensification require the current food production systems to go back to basics.

Principle 1. Circular Agriculture Should Be Regenerative

The reason we start with the principle of regeneration is because of the need to restore the health of the land, soils and water bodies that support food production. According to Hawken et al. (1999), soils are one of the natural capitals whose health needs to be preserved. Intensive agriculture has contributed to the degrading of the quality of soils. For example, a third of the topsoil in US has been lost (Hawken et al., 1999), the organic matter in the soils has also drastically diminished by 60% (The Nature Conservancy, 2016) leading to poor aeration, nutrient circulation and water retention. It is projected that within 60 years all of the world's topsoil could become unproductive if the current agricultural practices and soil degradation continue (Maximillian et al., 2019).

Circular agriculture should contribute towards restoring the health of the soils through practices, such as no-till farming and replacement of the use of synthetic fertilisers through leguminous plants or animal manure (Morseletto, 2020). Morseletto (2020) further distinguishes between restoration as an endless cycle of use, reuse and repair, and regeneration which upgrades and maintains the conditions of the ecosystems' functionality through self-renewal. It is the latter aspect that is particularly relevant to circular agriculture as it supports nature's ability to "revive itself, recover from disturbances, and rebuild its functions" (Morseletto, 2020, p. 7). In this context, there are many agricultural practices that allow for regeneration and can be adapted to different locations and parts of the world (Pretty, 1995). The list includes biodynamic farming (e.g. Osthaus, 2010), organic farming (e.g. Hansen, 2010), permaculture (e.g. Holmgren, 2002), regenerative farming (e.g. Massy, 2020) as well as methods based on advances in bioinformatics (Parray & Shameem, 2019).

Principle 2. Circular Agriculture Should Be Land-Effective

Land availability, and in particular land suitable for farming, is a physical constraint for a growing global population expected to reach 11 billion by 2100 (UN, 2019). Agriculture currently uses 34% of the landmass available on this planet compared to 1% occupied by human settlements, infrastructure and other built-up areas (Ritchie & Roser, 2019). Livestock, including livestock grazing and feed, uses disproportionately more land, namely 27% of the total landmass and 77% of total agricultural land, while all remaining food production is grown on 7% of the planet's landmass and 23% of agricultural land (Ritchie & Roser, 2019, see Fig. 5.1). Furthermore, the pressure from food production has left 25% of the global cultivated area degraded (Global Panel on Agriculture and Food Systems for Nutrition, 2020).

The second principle essentially says that circular agriculture should use as much land as necessary to achieve effective food production. Instead of thinking about production efficiencies which imply maximising outputs while minimising resources used, effectiveness aims at achieving the desired best result. While the emphasis in efficiency is on scarcity and maximising production, a major aspect of effectiveness is that a choice is made between heterogeneous activities and the different development potential resources have (Forbord, 2017). This is extremely important when analysing the potential of resources, such as land, soil, water or climatic conditions, to produce food. Put simply, effectiveness describes the extent to which the

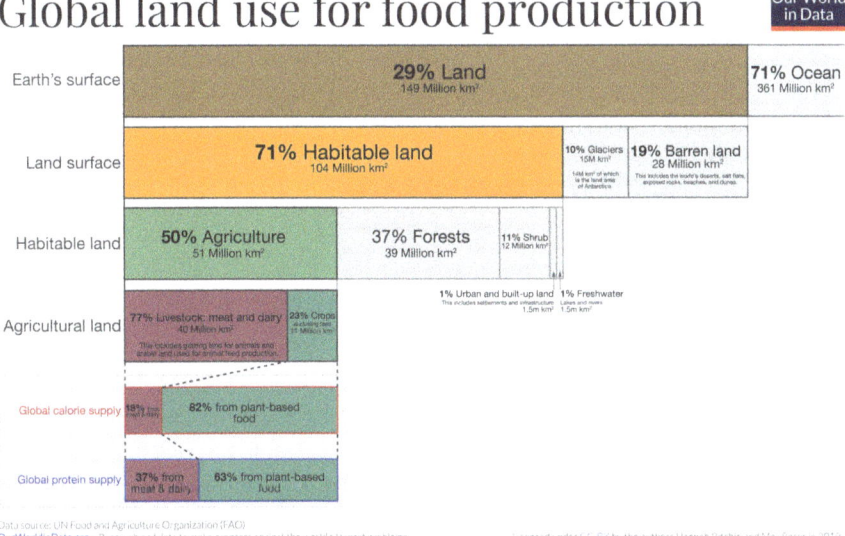

Source: Ritchie & Roser (2019)

Fig. 5.1 Global land use for food production. (Source: Ritchie & Roser, 2019)

objective of producing quality food is met rather than maximising food production (Productivity Commission, 2013). For centuries, the aim of intensive and industrial agriculture has been simply to produce more food without considering sustainability implications and this has ultimately contributed to the current planet emergency, as manifested among other things through deterioration of soil fertility, desertification and land contamination as well as climate change.

This principle calls for such exploitative and destructive practices to change by considering all possible alternatives. A good example is livestock and meat production. Although animal-based products can have a place in the human diet, their share needs to be drastically reduced as they require excessively more land. The EAT-Lancet Commission recommends a 50% global reduction in the current levels of red meat consumption (Willett et al., 2019). In cases, such as Australia and USA, this reduction needs to be 80–90%. Alternatives that are better for both, human and ecological health, are plant-based, including legumes, vegetables, whole grains and nuts, and they represent more effective use of the planet's landmass. Human diets across the globe need to shift towards providing adequate, safe and healthy nutrition while optimising the use of land and all other resources. They also need to be "protective and respectful of biodiversity and ecosystems, culturally acceptable, accessible, economically fair and affordable" (FAO, 2010, p. 33).

Smart land use and innovative food production technologies are also related to this principle. Examples include agroforestry, urban agriculture, vertical farming, hydroponics, use of moisture sensors, drip irrigation, innovative application of nitrogen-fixing bacteria and other microbe-based ways to improve the soils, indoor agriculture, growing underground, floating vegetable gardens, use of drones for planting and monitoring, robots for picking fruit and vegetables, blockchain technology for tracing provenance and losses. The list of innovative solutions is long and constantly expanding. Many of these ideas originate locally in countries, such as India (e.g. Agriculture & Food, 2020) and Bangladesh (e.g. Sunder, 2020) while other opportunities continue to emerge from multinational and technologically advanced companies (e.g. The Economist, 2016). They all potentially allow for different options to be explored and used to achieve land-efficiency in food production.

Principle 3. Circular Agriculture Requires Integrated Management

Integrated management requires all aspects and elements of agriculture to work together and synergistically towards the unified objective to respond to the planetary emergency while continuing to produce quality food. Particular issues that need to be considered are pest and weed management, the use of pesticides and herbicides, water efficiencies, crop selection, cover crops and rotation, and most importantly, agriculture should not result in harmful contaminants introduced into the soil, water bodies and the atmosphere.

Integrated pest and weed management is a process which controls and prevents the damage caused by harmful organisms "through a combination of techniques such as biological control, habitat manipulation, modification of cultural practices, and use of resistant varieties" (University of California, 2021, para. 2). It looks at eliminating the conditions that attract pests to avoid them becoming a problem. However, the actual terms "pest" and "weed" are human-centred and a myriad of organisms (e.g. insects, molluscs, nematodes), microorganisms (e.g. bacteria, fungi) and plants which have grown at the wrong place, are blamed to be harmful entirely from the point of view of food production while they have a legitimate place in the planet's ecosystems (Flint & van den Bosch, 2012). A lot of human effort has gone into creating chemical-based ways to control pests and weeds. In most cases, this has resulted in negative side effects and development of long-term resistance in the targeted species. Rachel Carson's *Silent Spring* (1962) which revealed the harm done by the now infamous DDT, was the starting point of the global environmental movement. This work was then continued by Vandana Shiva (1989) and many others. Their message is that pest and weed management is an ecological issue (Flint & van den Bosch, 2012) and it should be handled by relying on natural factors, such as weather patterns and normal biological enemies. While monitoring and information gathering are important, the solutions should be found with the understanding that food production is part of the larger biosphere on this planet and any human actions would have implications for the natural environment.

Crop rotation is a well-known method to maintain soil fertility while continuing to produce quality food. Legumes (peas, beans, soybeans, lentils and others) are usually some of the plants used because of their ability to fix nitrogen in the soil. The rhizobacteria in the legumes' nodules convert the nitrogen from the air into a form usable by plants, which improves the fertility of the soil and reduces the need for fertilisers. Legumes are also water-efficient – a very important consideration in dry climates, such as in Australia. Annual rotation of legumes with other crops provides good results. For example, a 15-year study in USA showed that rotation between corn and soybeans increased corn yields by 20% while nitrous oxide emissions (which have 284 times higher global warming potential than CO_2) were reduced by 35%; soybeans yields were 7% higher with nitrous oxide emissions not affected (Behnke et al., 2018). A meta-analysis of 45 studies with 214 comparisons of crop rotations in China (Zhao et al., 2020) showed significant benefits – 20% increased yields compared to monocultures with the effects being the biggest when legumes were the pre-crops (yields increased by 27%). Crop rotation helps break the resistance of the western corn rootworm – a major pest in many American states, even in the case of genetically engineered corn which was supposed not to be affected by it (Carrière et al., 2020). Similar outcomes are found in India, Africa, Europe and across the world with crop rotation seen as a way to sustainably intensify agriculture (e.g. Kumar et al., 2020) and improve its resilience to climate change (Bowles et al., 2020). Hence, within a circular-agriculture model, crop rotation can be a powerful way to improve yields and also mitigate weed, pest and pathogen pressure as well as maintain plant diversity.

Integrated management in circular agriculture also implies the elimination of contaminants. They range from chemicals, such as the ones applied as fertilisers, pesticides or herbicides, to plastics linked predominantly with packaging used during food production, and microplastics – tiny particles produced from the breaking down of plastics, as well as with bio-solids from treated waste-water sludge that ends up on agricultural lands. Some of the contaminants associated with bio-solids are microplastics, nanoplastics, synthetic materials which do not naturally decompose, heavy metals, pharmaceuticals and engineered nanoparticles (Mohajerani & Karabatak, 2020). It is ironic that a lot of other waste-handling solutions aimed at a circular economy create longer-term problems for agricultural lands. Hence, it is important to keep circular agriculture as a separate, highly important field of human endeavour and protect the land from being contaminated. There is however a major group of contaminants associated with the use of antimicrobials and antibiotics for farm animals to prevent infections or promote growth, which ultimately find their way in the water bodies and on the land (Review on Antimicrobial Resistance, 2016).

Given the importance of food and the land on which it is grown, environmental health becomes a top priority and agricultural soils should not be used as a dumping ground for waste and effluents generated elsewhere. Circular agriculture itself should only use processes that are ecologically acceptable and part of the natural cycle of elements.

Principle 4. Circular Agriculture Should Focus on Quality Not Quantity of Food Produced

Food production ultimately provides nutrition. There are vast differences in how people satisfy their hunger. For example, despite their popularity, fast food outlets are notorious for offering unhealthy food options. Diets rich in red meat are also considered unhealthy because of established correlations with cancers, and colorectal cancer in particular (WHO, 2015). Circular agriculture should thus focus on producing quality food, rather than large quantities and cheap options. Almost in all countries across the globe, people will benefit from eating diverse plant-based options, including staple foods such as rice, cereal grains, starchy vegetables and tubers as well as vegetables, wholegrains, pulses, nuts and fruits, which are nutrient-rich and do not compromise human health (Global Panel on Agriculture and Food Systems for Nutrition, 2020).

The definition adopted by the Global Panel on Agriculture and Food Systems for Nutrition (2016, p. 32), namely sustainable diets: "are those that eliminate hunger, are safe, reduce all forms of malnutrition, promote health and are produced sustainably, i.e. without undermining the environmental basis to generate high-quality diets for future generations", should be a guiding principle for circular agriculture. Years of intensive farming focused on increased yields have locked food systems "in a spiral of decline with environmental systems" (Global Panel on Agriculture and Food Systems for Nutrition, 2020, p. 17). In essence, the current food systems are failing to deliver adequate nutrition and have compromised the ecological environment. Adopting circular agriculture allows for systematic change to occur which will open up a lot of opportunities to respond to the planetary emergency by improving the health of all – people, soils, plants and many native species. Applications of new technologies can allow for precision agriculture within a circular model which reduces the need for external inputs and also for harvesting at the best moment maximising the nutritional content.

Many weaknesses and the fragility of human food systems were revealed during the COVID-19 pandemic with disrupting production and distribution chains. In Australia, for example, where the coronavirus had a relatively low health impact, the closure of international borders prevented foreign seasonal workers from participation in harvesting and fruit collection exposing the dependence of the country's agriculture on outside labour and lack of self-reliance. This was also a consequence from the preference for large intensive broad-acre agricultural properties with monocultures, a lot of them destined for export, where the aim was to produce large quantities rather than nutritional crop diversity. Circular agricultural systems are more resilient as they are self-reliant, bridge the gap between human and ecological health and provide not only food, but also jobs, support livelihoods and deliver ecological services.

Principle 5. Circular Agriculture Should Be Localised

These days it is difficult to envisage food production without international trade. Local food crop production can currently fulfil the demand for less than a third of the global population (Kinnunen et al., 2020).

Although circular agriculture does not impose being self-sufficient, it puts a large emphasis on localisation. It implies the choice of varieties which are best suited to the locality, the use of local seeds and traditional knowledge (Food & Business Knowledge Platform, 2019). Furthermore, any technological solutions need to be localised depending on geographic conditions, climate, availability of resources and cultural traditions. Smallholder farms and local farmers' markets have a much more pronounced importance as they are better suited to respond to the need for diversification and self-reliance (Global Panel on Agriculture and Food Systems for Nutrition, 2020). Feedback economic and policy mechanisms, such as based on taxes and subsidies, are easier to implement. As far as labour force on the farm, even the current agricultural systems rely heavily on the farmer's family plus machinery (Forbord & Vik, 2017) and occasional seasonal workers. In a circular agriculture, local workers and employment opportunities are essential.

Localisation also implies proximity to consumers which eliminates unnecessary food miles and storage while improving the freshness and nutritional value of the produce. However, it is important to keep in mind that although the distance food travels contributes towards its environmental footprint, the most effective way to reduce the planetary impacts of our nutritional choices is by eliminating or decreasing the consumption of animal-based products (Leavens, 2017).

Following the five principles described above, the food production systems will need to be redesigned as circular agriculture which produces valuable outputs, minimises their impacts on the environment and sustains both, the people and the productivity of the land. Circular agriculture also embraces new technological solutions and is capable to respond and adapt to changing social and environmental conditions. People's values and nutritional preferences can shape the concrete solutions but what is a dominant characteristic is the way waste is eliminated or absorbed in the system.

Waste Products and Food Production

Waste – another human-centred term, is the undesirable product at the end of the linear production model which circular agriculture has to eliminate. The aim of circular agriculture is to eliminate any wastage and reintegrate everything within a biochemical cycle. For the time being however, there are at least three different aspects of waste products related to food production that need to be understood, namely: pollution associated with agricultural processes which covers contamination of the soils, waters and atmosphere; waste from other sectors of the economy deposited on agricultural land; and lost and unwanted food or produced residues. Let us look at all three of them in turn.

Pollution Generated by Agriculture

Agriculture contributes largely to water and land pollution by discharging nutrients, pesticides, salts, sediments, organic carbon, pathogens, heavy metals and drug residues and it causes 70% of the obstructions of waterways (Mateo-Sagasta et al., 2017). The overuse of fertilisers is responsible for eutrophication of lakes and waterbodies (UNDP, 2016). Pressure on fresh waterbodies and marine environments also comes from the intensive livestock sector responsible not only for manure and excreta discharge but also for contamination with effluents from slaughter houses. Antibiotics, growth hormones and vaccines applied to farm animals are further exacerbating the problems with water pollution. This creates conditions for the emergence of zoonotic waterborne pathogens and other ill effects on human health (Mateo-Sagasta et al., 2017).

Adopting circular agriculture would put a stop on such unhealthy discharges and prevent the pending water crises. This will also reduce the amounts of water withdrawal and allow for regeneration of rivers, lakes and swamp areas together with the species which inhabit them. In Australia, for example, the Murray-Darling Basin which is home of 16 internationally significant wetlands and 35 endangered species

(Murray-Darling Basin Authority, n.d.) has been under tremendous pressure from agriculture and the withdrawal of fresh water has caused mass fish deaths. Agricultural run-off is one of the biggest threats to the Great Barrier Reef which combined with climate change making bleak the future of this natural heritage of global importance. A major cause for the run-off is the large-scale deforestation and clearing of native vegetation for livestock pasture and grazing.

Waste from Other Economic Activities

Increasing the level of organic material in the soils is a natural way to store carbon. Agriculture is therefore seen as a solution to manage wastes from other economic activities. There is a strong argument that agriculture can not only produce food but also deliver ecological services through carbon sequestration: "The carbon that is removed from the atmosphere and captured in soils and plant biomass is the same carbon that makes agricultural soils more fertile" (World Bank, 2012, p. ix). The biochemical process that allows this is photosynthesis. It captures light energy and uses it to convert carbon dioxide, water and minerals into oxygen needed by all living organisms and organic material (Britannica, 2021). Soil carbon improves the fertility of the land as it helps hold and release water and other nutrients. New technologies, including sensors and robots, are being deployed to measure and monitor the carbon content in soils.

Biochar – a solid, carbon-rich product or charcoal produced from biomass through pyrolysis in the absence of oxygen, specifically for application to soil (Hyland & Sarmah, 2014), is seen as a promising technological solution. It can be produced using household and industrial wastes (e.g. food scraps and sludges) as well as from agricultural plant material (e.g. from pruning) (Hou & O'Connor, 2020). Research has shown a lot of benefits from using this material to improve soil fertility, remediate contaminants, increase crop production and most importantly, sequester carbon for centuries (Jindo et al., 2020). Biochar can also be used as a growing medium and can be mixed with compost. Whether this product has a role to play within circular agriculture however largely depends on many specific factors. Amongst them is the quality of the feedstock for the pyrolysis, e.g. whether it contains any heavy metals, salts, chemical residues or other contaminants and the type of energy used, preferably renewable. Negative changes in soil properties may also occur because of decreased availability of nitrogen, higher than optimal pH or changes in the mineralisation processes (Jindo et al., 2020). Furthermore, the pyrolysis (which can be done in small-scale and home-made devices) is not always cost-effective (Kuppusamy et al., 2016) and in addition to bio-oil, also produces syngas requiring careful handling. Despite of potential benefits, the review of the latest scientific evidence concludes that: "A proper combination among biochar type, the purpose of its use and optimum application rate should be explored. Selecting the right biomass and optimisation of pyrolysis conditions are key factors to tailor high agronomic value biochar for its proper use in the agricultural domain" (Jindo et al., p. 8).

The easiest way to maintain carbon sequestration is to preserve natural vegetation in old-growth forests, steppes, savannahs and wetlands by preventing further land clearing for agricultural purposes. This is one of the targets set by Greenpeace (2018). A circular economy can combine agroforestry to prevent further destruction of natural habitat. Reducing erosion, applying no-tillage methods, avoiding residue removal or burning and drainage of peat lands are all possible methods to look after the soil in circular agriculture.

Food Loss and Residues

The Food and Agriculture Organisation of the United Nations (FAO, 2019) distinguished between food loss – occurring during the production and supply chain excluding retail operations estimated at 13.8% of all food produced, and food waste – taking place at the level of retailing and consumption. Despite the significance of food waste, estimated at 19.2% (or a total of 33% of food loss and waste according to FAO, 2013) and with calls for it to be halved significantly (SDG Target 12.3; also Willett et al., 2019), only food loss is of interest from a circular agriculture perspective. Another aspect of interest are residues and materials that do not represent food or feed but are an integral part of agricultural production, e.g. inedible parts.

There are many reasons for food losses. They are generally linked to unintended circumstances in the agricultural processes or technical limitations. In some cases, food is left unharvested because of weather conditions, diseases, poor quality, lack of labour, infrastructure, equipment or storage facilities or because of unfavourable economic reasons. Other reasons are linked to the way food is stored or transported with spillages possible. Also, not all produced food is considered good for the market. Sometimes the reasons to reject part of the harvest may be purely cosmetic (e.g. the wrong shape, size or physical damage) and not linked to the nutritional value of the produce. Within a circular economy, food loss is not considered waste as the products could either still be marketed, be it at a lower price, or apply improved technologies to avoid it altogether. Furthermore, such unwanted or damaged food can be returned as organic fertiliser back into the soil. Sometimes the latter is described as valorising or creating further value (Muscio & Sisto, 2020) while in essence it assists the circulation of nutrients.

The residual biomass produced in circular agriculture, such as leaves, stems, husks and crop remnants, can be used in some cases as animal feed but also as natural bio-fertilisers to maintain the fertility of the soils (Wageningen University & Research, 2018). Manure (animal and domestic) and compost can also be used as can biochar. A major consideration in relation to livestock agriculture is restricting its use of land which also means that farm animals should not be fed grains and plant-based proteins that are suitable for human consumption, for example corn or soybeans. Feeding grains to livestock so that humans then consume animal-based foods is a very inefficient way of providing nutrients to people, the least efficient option being beef (Eshel et al., 2014; Clark & Tilman, 2017; Poore & Nemecek, 2018). In fact, from a moral human perspective, this is one of the largest wastes of calories in food production whilst also being the highest contributor to land, air and water pollution.

Implementing Circular Agriculture

One of the earliest schools of economic thought – the physiocrats in the eighteenth century, described agriculture as being the source of all wealth. The physiocrats also saw agriculture as a circular flow of money similar to the blood system (Murray et al., 2017). Industrialisation strengthened further the concept about money circulating within the economy and the market-based capitalist systems took this to perfection. However, what was left outside the economic closed-loop model was the costs of the services provided by the biophysical environment and the costs of repairing this natural capital in order to continue to provide such functions. Environmental economics (Daly & Farey, 2004; Thampapillai & Sinden, 2013) tried to re-integrate the natural environment in the economic model of money circulation.

The implementation of a circular economy and circular agriculture however is not about the flow of money but about the use of resources and biochemical cycles on this planet, almost all of which have been impacted by human activities. From a sustainability perspective, circular agriculture also needs to simultaneously maximise ecosystem functioning and human well-being (Murray et al., 2017). In fact, the main challenge in implementing circular agriculture is that the economic dimension needs to be subordinated to the ability of the land to produce food and provide quality nutrition. Such a transition requires not only adjustment in the way we think about agriculture but also strong policy support.

It seems that it is easier to make some progress in other sectors in the economy than in agriculture. Circular economy examples are not only the Legacy Living Lab in Fremantle but also industrial ecology parks, such as the Kwinana Industrial Area in Western Australia (van Beers et al., 2008). They incorporate significant innovations and a major shift in the way of thinking about production and construction processes through prolonging the life cycle of the products and closing industrial loops where the outputs from one manufacturer become inputs for another. Research and innovation have historically played a significant role in agriculture and pushed high the levels of productivity. For a circular agriculture, technological progress is also very important but alone, it is not enough as similarly there need to be changes in the way of thinking. An analysis of the scale of changes required in the European Union describes the challenges as "daunting" and requiring significant government support without a clear understanding about the size of the necessary transformations (Muscio & Sisto, 2020). What is clear however is that the planetary emergency is offering a narrow window of opportunity as agriculture is particularly vulnerable to climate change with resulting increasing temperatures, higher frequencies of extreme weather events and rainfall variations. Circular agriculture needs to rely on research, new digital technologies, artificial intelligence and other innovations to become climate-resilient.

A transition to the new circular model of food production is a complex process that involves all sectors of society and its institutions. Government legislation and policies need to facilitate such a shift and despite broad acknowledgement of the severity of the situation, only a limited number of countries have started to facilitate such a transition. Examples of countries which have started to legislate circular agriculture include the Netherlands and China. Other countries, such as Australia and most European Union members, have policies that specifically target food waste with less progress made on the importance of circularity. Europe, including UK, dominates research in the area of circular agriculture producing 80% of publications in this area (Esposito et al., 2020). Interest in circular agriculture for improved food security is also present in Sub-Saharan Africa (Boon & Anuga, 2020). Possible demonstration models emerge – "lighthouse farms" (Wageningen University & Research, 2018) and inspirational places (e.g. Fair Harvest, 2021 in Western Australia). Their models may be unique but they prove what is possible.

The World Economic Forum (2000) identifies the need to realign incentives within each economy to produce food that is healthy for the people and the planet. This involves investment from the government, innovative business models, investments that target environmental and social benefits alongside financial returns and most importantly, changes in consumer preferences towards foods that are nutritious while being produced in an environmental and socially responsible way. The circular agriculture model can potentially satisfy all these requirements and it needs to be seen as the top priority of this century.

Conclusion

The planetary emergency is essentially interlinked with the current agricultural systems: "The climate crisis, soil degradation, rising ocean levels, biodiversity loss, pollution of air, water and land, and depletion of freshwater resources all pose risks to, and are partly driven by, the way food systems work" (Global Panel on Agriculture and Food Systems for Nutrition, 2020, p. 51). Circular agriculture offers most-needed solutions to dealing with this grave situation. Food production heavily depends on maintaining the health of soils. Its five principles guarantee that agriculture will be regenerative, use land in a most effective way, will be managed in an integrated way with a focus on the quality (and not quantity) of the produced food and with a strong localisation to serve consumers, building on their local skills and knowledge. The human concept of waste will be truly eliminated – nothing produced within a circular agriculture would at any point become waste (World Economic Forum, 2020).

Circular agriculture itself and transitioning to it will have different models and pathways. They will rely on innovation and policy support as well as on consumers' ability and desire to shift their expectations towards nutritional and environmentally and socially responsible food choices. Humans and their food requirements will have to be part of the biological cycles of the planet with minimal generation of waste and elimination of contaminants. This is a complex socio-technical transformation in which nature re-emerges as the main resource that needs to be protected and maintained.

References

Agriculture & Food. (2020). *e-Newsletter*. Volume 2, Issue 6. http://www.agrifoodmagazine.co.in/wp-content/uploads/2020/05/Volume-2-Issue-6-June-2020.pdf

Behnke, G. D., Zuber, S. M., Pittelkow, C. M., Nafziger, E. D., & Villamil, M. B. (2018). Long-term crop rotation and tillage effects on soil greenhouse gas emissions and crop production in Illinois, USA. *Agriculture, Ecosystems & Environment, 261*, 62–70. https://doi.org/10.1016/j.agee.2018.03.007

Boon, E. K., & Anuga, S. W. (2020). Circular economy and its relevance for improving food and nutrition security in Sub-Saharan Africa: The case of Ghana. *Materials Circular Economy, 2*, 5. https://doi.org/10.1007/s42824-020-00005-z

Bowles, T. M., Mooshammer, M., Socolar, Y., Calderón, F., Cavigelli, M. A., Culman, S. W., … Grandy, A. S. (2020). Long-term evidence shows that crop-rotation diversification increases agricultural resilience to adverse growing conditions in North America. *One Earth, 2*(3), 284–293. https://doi.org/10.1016/j.oneear.2020.02.007

Britannica. (2021). *Photosynthesis*. https://www.britannica.com/science/photosynthesis

Carrière, Y., Brown, Z., Aglasan, S., Dutilleul, P., Carroll, M., Head, G., … Carroll, S. P. (2020). Crop rotation mitigates impacts of corn rootworm resistance to transgenic Bt corn. *Proceedings of the National Academy of Sciences (PNAS), 117*(31), 18385–18392. https://doi.org/10.1073/pnas.2003604117

Carson, R. (1962). *Silent spring*. Houghton Mifflin Harcourt.

Clark, M., & Tilman, D. (2017). Comparative analysis of environmental impacts of agricultural production systems, agricultural input efficiency, and food choice. *Environmental Research Letters, 12*, 064016.

Crippa, M., Solazzo, E., Guizzardi, D., Monforti-Ferrario, F., Tubiello, F. N., & Leip, A. (2021). Food systems are responsible for a third of global anthropogenic GHG emissions. *Nature Food, 2*, 198–209. https://doi.org/10.1038/s43016-021-00225-9

Curtin University. (n.d.). *Legacy Living Lab*. https://l3.curtin.edu.au/about/

Daly, H. E., & Farley, J. (2004). *Ecological economics: Principles and applications*. Island Press.

Ellen MacArthur Foundation. (2017). *Food and the circular economy*. https://www.ellenmacarthurfoundation.org/explore/food-cities-the-circular-economy

Eshel, G., Shepon, A., Makov, T., & Milo, R. (2014). Land, irrigation water, green- house gas, and reactive nitrogen burdens of meat, eggs, and dairy production in the United States. *Proceedings of the National Academy of Sciences of the United States of America (PNAS), 111*(33), 11996–12001. https://doi.org/10.1073/pnas.1402183111

Esposito, B., Sessa, M. R., Sica, D., & Malandrino, O. (2020). Towards circular economy in the agri-food sector: A systematic literature review. *Sustainability, 12*(18), 7401. https://doi.org/10.3390/su12187401

Fair Harvest. (2021). *Permaculture*. https://www.fairharvest.com.au

Fan, W., Dong, X., Wei, H., Weng, B., Liang, L., Xu, Z., … Song, C. (2018). Is it true that the longer the extended industrial chain, the better the circular agriculture? A case study of circular agriculture industry company in Fuqing, Fujian. *Journal of Cleaner Production, 189*, 718–728. https://doi.org/10.1016/j.jclepro.2018.04.119

Flint, M. L., & van den Bosch, R. (2012). *Introduction to integrated pest management*. E-Book, SpringerLink.

Food & Business Knowledge Platform. (2019). *Circular agriculture in low and middle income countries*. https://knowledge4food.net/wp-content/uploads/2020/03/191016_fbkp-circular-agriculture-lmics_discussionpaper.pdf

Food and Agriculture Organisation of the United Nations (FAO). (2010). *Sustainable diets and biodiversity: Directions and solutions for policy, research and action*. http://www.fao.org/3/i3004e/i3004e.pdf

Food and Agriculture Organization of the United Nations (FAO). (2013). *Food wastage footprint: Impacts on natural resources*. Summary Report. http://www.fao.org/3/i3347e/i3347e.pdf

Food and Agriculture Organization of the United Nations (FAO). (2019). *The state of food and agriculture, moving forward on food loss and waste reduction*. http://www.fao.org/3/ca6030en/ca6030en.pdf

Forbord, M. (2017). *Efficiency and effectiveness in agricultural related activity patterns*. https://www.impgroup.org/uploads/papers/57.pdf

Forbord, M., & Vik, J. (2017). Food, farmers, and the future: Investigating prospects of increased food production within a national context. *Land Use Policy, 67*, 546–557. https://doi.org/10.1016/j.landusepol.2017.06.031

Global Panel on Agriculture and Food Systems for Nutrition. (2016). *Food systems and diets: Facing the challenges of the 21st century*. https://www.glopan.org/foresight1/

Global Panel on Agriculture and Food Systems for Nutrition. (2020). *Future food systems: For people, our planet, and prosperity*. Foresight 2.0. https://www.glopan.org/foresight2/

Greenpeace. (2018). *Less is more: Reducing meat and dairy for a healthier life and planet. The Greenpeace vision of the meat and dairy system towards 2050*. https://www.greenpeace.org/static/planet4-international-stateless/2018/03/698c4c4a-summary_greenpeace-livestock-vision-towards-2050.pdf

Hansen, A. L. (2010). *The organic farming manual: A comprehensive guide to starting and running a certified organic farm*. Storey Publishing.

Hawken, P., Lovins, H., & Lovins, A. (1999). *Natural capitalism: Creating the next industrial revolution*. Little, Brown & Company.

Holmgren, D. (2002). *Permaculture principles and pathways beyond sustainability*. Holmgren Design Services.

Hou, D., & O'Connor, D. (2020). Green and sustainable remediation: Concepts, principles, and pertaining research. In D. Hou (Ed.), *Sustainable remediation of contaminated soil and groundwater: Materials, processes, and assessment* (pp. 1–17). Butterworth-Heinemann.

Hyland, C., & Sarmah, A. K. (2014). Advances and innovations in biochar production and utilisation for improving environmental quality. In V. K. Gupta, M. G. Tuohy, C. P. Kubicek, J. Saddler, & F. Xu (Eds.), *Bioenergy research: Advances and applications* (pp. 435–446). Elsevier.

Jindo, K., Sánchez-Monedero, M. A., Mastrolonardo, G., Audette, Y., Higashikawa, F. S., Silva, C. A., … Mondini, C. (2020). Role of biochar in promoting circular economy in the agriculture sector. Part 2: A review of the biochar roles in growing media, composting and as soil amendment. *Chemical and Biological Technologies in Agriculture, 7*, 16. https://doi.org/10.1186/s40538-020-00179-3

Kinnunen, P., Guillaume, J. H. A., Taka, M., D'Odorico, P., Siebert, S., Puma, M. J., Jalava, M., & Kummu, M. (2020). Local food crop production can fulfil demand for less than one-third of the population. *Nature Food, 1*, 229–237. https://doi.org/10.1038/s43016-020-0060-7

Kumar, R., Mishra, J. S., Rao, K. K., Mondal, S., Hazra, K. K., Choudhary, J. S., … Bhatt, B. P. (2020). Crop rotation and tillage management options for sustainable intensification of rice-fallow agro-ecosystem in eastern India. *Scientific Reports, 10*, 11146. https://doi. org/10.1038/s41598-020-67973-9

Kuppusamy, S., Thavamani, P., Megharaj, M., Venkateswarlu, K., & Naidu, R. (2016). Agronomic and remedial benefits and risks of applying biochar to soil: Current knowledge and future research directions. *Environment International, 87*, 1–12. https://doi.org/10.1016/j. envint.2015.10.018

Leavens, M. (2017). *Do food miles really matter?* Harvard University. https://green.harvard.edu/ news/do-food-miles-really-matter

Massy, C. (2020). *Call of the reed warbler: A new agriculture, a new earth* (revised and updated ed.). University of Queensland Press.

Mateo-Sagasta, J., Zadeh, S. M., Turral, H., & Burke, J. (2017). *Water pollution from agriculture: A global review*. Food and Agriculture Organisation of the United Nations, Rome. http://www. fao.org/3/a-i7754e.pdf

Maximillian, J., Brusseau, M. L., Glenn, E. P., & Matthias, A. D. (2019). Pollution and environmental perturbations in the global system. In M. L. Brusseau, I. L. Pepper, & C. P. Gerba (Eds.), *Environmental and pollution science* (3rd ed., pp. 457–476). Academic.

Mohajerani, A., & Karabatak, B. (2020). Microplastics and pollutants in biosolids have contaminated agricultural soils: An analytical study and a proposal to cease the use of biosolids in farmlands and utilise them in sustainable bricks. *Waste Management, 107*, 252–265. https:// doi.org/10.1016/j.wasman.2020.04.021

Morseletto, P. (2020). Restorative and regenerative: Exploring the concepts in the circular economy. *Journal of Industrial Ecology, 24*(4), 763–773. https://doi.org/10.1111/jiec.12987

Murray, A., Skene, K., & Haynes, K. (2017). The circular economy: An interdisciplinary exploration of the concept and application in a global context. *Journal of Business Ethics, 140*, 369–380. https://doi.org/10.1007/s10551-015-2693-2

Murray-Darling Basin Authority. (n.d.). *The Murray–Darling Basin and why it's important.* https://www.mdba.gov.au/importance-murray-darling-basin

Muscio, A., & Sisto, R. (2020). Are agri-food systems really switching to a circular economy model? Implications for European research and innovation policy. *Sustainability, 12*, 5554. https://doi.org/10.3390/su12145554

Organisation for Economic Co-operation and Development (OECD). (2015). *The metropolitan century: Understanding urbanisation and its consequences*. Policy highlights. https://www. oecd.org/regional/regional-policy/The-Metropolitan-Century-Policy-Highlights%20.pdf

Osthaus, K.-E. (2010). *The biodynamic farm: Developing a holistic organism*. Floris Books.

Parray, J. A., & Shameem, N. (2019). *Sustainable agriculture: Advances in plant metabolome and microbiome*. Academic.

Pelletier, N., & Tyedmers, P. (2010). Forecasting potential global environmental costs of livestock production 2000–2050. *Proceedings of the National Academy of Sciences of the United States of America (PNAS), 107*(43), 18371–18374. https://doi.org/10.1073/pnas.1004659107

Poore, J., & Nemecek, T. (2018). Reducing food's environmental impacts through producers and consumers. *Science, 360*, 987–992. https://doi.org/10.1126/science.aaq0216

Pretty, J. N. (1995). *Regenerating agriculture: Policies and practice for sustainability and self-reliance*. Earthscan.

Productivity Commission. (2013). *On efficiency and effectiveness: Some definitions*. Australian Government. https://www.pc.gov.au/research/supporting/efficiency-effectiveness/efficiency-effectiveness.pdf

Review on Antimicrobial Resistance. (2016). *Tackling drug-resistant infections globally: Final report and recommendations*. https://amr-review.org/sites/default/files/160525_Final%20 paper_with%20cover.pdf

Ritchie, H., & Roser, M. (2019). *Land use*. Our world in data. https://ourworldindata.org/land-use

Schouten, C. (2020). *Circular agriculture: A vision for sustainability*. https://www.ifpri.org/blog/ circular-agriculture-vision-sustainability

Shiva, V. (1989). *The violence of the green revolution: Ecological degradation and political con-flict in Punjab*. Natraj Publishers.

Smith, P., House, J. I., Bustamante, M., Sobocká, J., Harper, R., Pan, G., … Pug, T. A. M. (2016). Global change pressures on soils from land use and management. *Global Change Biology, 22*(3), 1008–1028. https://doi.org/10.1111/gcb.13068

Sunder, K. (2020). *The remarkable floating gardens of Bangladesh*. https://www.bbc.com/future/article/20200910-the-remarkable-floating-gardens-of-bangladesh

Thampapillai, D. J., & Sinden, J. A. (2013). *Environmental economics: Concepts, methods, and policies*. Oxford University Press.

The Economist. (2016). *Technology quarterly: The future of agriculture*. https://www.economist.com/technology-quarterly/2016-06-09/factory-fresh

The Nature Conservancy. (2016). *reThink soil: A roadmap to US soil health*. https://www.nature.org/content/dam/tnc/nature/en/documents/rethink-soil-executive-summary.pdf

Timmis, K., & Ramos, J. L. (2021). The soil crisis: The need to treat as a global health problem and the pivotal role of microbes in prophylaxis and therapy. *Microbial Biotechnology, 14*(3), 769–797. https://doi.org/10.1111/1751-7915.13771

Toop, T. A., Ward, A., Oldfield, T., Hull, M., Kirby, M. E., & Theodorou, M. (2017). AgroCycle – Developing a circular economy in agriculture. *Energy Procedia, 123*, 76–80. https://doi.org/10.1016/j.egypro.2017.07.269

United Nations (UN). (2019). *World population prospects 2019: Highlights*. Department of Economic and Social Affairs, Population Division. ST/ESA/SER.A/423. https://population.un.org/wpp/Publications/Files/WPP2019_Highlights.pdf

United Nations Development Programme (UNDP). (2016). *A Snapshot of the world's water quality: Towards a global assessment*. Nairobi, Kenya. https://uneplive.unep.org/media/docs/assessments/unep_wwqa_report_web.pdf

University of California. (2021). *What is integrated pest management (IPM)?* https://www2.ipm.ucanr.edu/What-is-IPM/

Van Beers, D., Bossilkov, A., Corder, G., & van Berkel, R. (2008). Industrial symbiosis in the Australian minerals industry: The cases of Kwinana and Gladstone. *Journal of Industrial Ecology, 11*(1), 55–72. https://doi.org/10.1162/jiec.2007.1161

Wageningen University & Research. (2018). *Circular agriculture: A new perspective for Dutch agriculture*. https://www.wur.nl/en/newsarticle/Circular-agriculture-a-new-perspective-for-Dutch-agriculture-1.htm

Willett, W., Rockström, J., Loken, B., Springmann, M., Lang, T., Vermeulen, S., … Murray, C. J. L. (2019). Food in the Anthropocene: The EAT–Lancet Commission on healthy diets from sustainable food systems. *The Lancet, 393*(10170), 447–492. https://doi.org/10.1016/S0140-6736(18)31788-4

World Bank. (2012). *Carbon sequestration in agricultural soils*. https://openknowledge.worldbank.org/handle/10986/11868

World Bank. (2021). *Urban population (% of total population) – Australia*. https://data.worldbank.org/indicator/SP.URB.TOTL.IN.ZS?locations=AU

World Economic Forum. (2000). *Incentivising food systems transformation*. http://www3.weforum.org/docs/WEF_Incentivizing_Food_Systems_Transformation.pdf

World Economic Forum. (2020). *The circular economy*. https://olc.worldbank.org/content/circular-economy

World Health Organisation (WHO). (2015). *Cancer: Carcinogenicity of the consumption of red meat and processed meat*. https://www.who.int/news-room/q-a-detail/cancer-carcinogenicity-of-the-consumption-of-red-meat-and-processed-meat

Zaman, A., & Ahsan, T. (2019). *Zero-waste: Reconsidering waste management for the future*. Routledge.

Zhao, J., Yang, Y., Zhang, K., Jeong, J., Zeng, Z., & Zang, H. (2020). Does crop rotation yield more in China? A meta-analysis. *Field Crops Research, 245*, 107659. https://doi.org/10.1016/j.fcr.2019.107659

Chapter 6
Sustainability Transitions in Food Production

Abstract The human response in defying situations, such as the present planetary emergency combined with the COVID-19 pandemic, is to find technological solutions allowing society to transition on a new trajectory that transforms the current practices for good. The chapter outlines the need to transition to better food production practices and introduces new technological changes related to vertical farming, artificial constructions (or ponics), agroecology, agroforestry, the use of drones and nanotechnology and application of ICTs, including artificial intelligence, 3D printing and blockchain. This facilitates clever food production and choices that are environmentally and nutritionally better.

Back in 2018, in a spontaneous conversation about our world food systems with a small-farm vegetables grower near Philadelphia, in the United States, the farmer shared what he described as sad observations. The trends he outlined clearly spoke about the disappearance of small farms and the utter dominance of industrial agricultural mass production based on land consolidation into large properties under the ownership of rich and powerful people who were happy with their conveyor-type bulk food operations. Individual small players like him had only little chance to survive. It was not only because of the rise of farm consolidation and the thriving of industrialised factory farming dominated by technology and scale efficiencies in production, but also due to the collapsing of commodity prices related to globalisation and trade wars. The farmer also added climate change-associated severe weather conditions and political polarisation to the long list of factors that were threatening his livelihood. There was no doubt that the advent of technology was making farms more efficient, but the new economic benefits from technological development were going to large farms at the expense of small farmers like him who were forced to sell off their land. The farmer also stressed that the actions of all of these large farms focussed on constantly increasing production, were having detrimental consequences for the natural environment. Such threat posed by agriculture to ecological systems was not yet fully recognised.

Many small farmers like him were only a step away from a consolidation destiny to be trapped in large-scale commercial farming. He however managed to get away from it finding a niche market as organic vegetables grower supplying selected restaurants around Philadelphia and in the Washington DC area. Many of his concerns resonated with us thinking about farming in Australia. He was talking also about the disappearance of hog, chicken and other small farms in USA and the unsustainable practices of growing the same crop year after year exhausting the soil instead of using crop rotation to help the land preserve its own resources and flourish. The big American farms, he said, shamelessly use genetically modified (GM) crop varieties and large amounts of chemical pesticides and fertilisers that damage the soils, water and air. They feed their livestock animals on feedlots using hormones to promote growth and fattening, and antibiotics to avoid the spread of infections. We asked the farmer how he sees the future and the answer he gave was similarly pessimistic. He plainly stated that the food system was not built to last because it squanders and degrades the resources on which it depends.

We already explained in previous chapters what the impacts of the current food production systems are on the planet's climate and ecological environment. Food is and will always remain a high priority for any species inhabiting the Earth. On the international climate policy agenda, food only appeared at the 23rd Conference of the Parties (COP23) to the United Nations Framework Convention on Climate Change (UNFCC) held in Bonn, Germany in 2017, which acknowledged the critical role agriculture plays in achieving the Paris Agreement. It was a landmark decision which on one hand recognised that the current practices are contributing to global warming and on the other, stressed the importance of innovation and climate-smart approaches to produce food (FAO, 2017b). By comparison, food security has been part of the sustainability journey from its inception with Agenda 21 (adopted in 1992), the Millennium Development Goals (2000–2015) and the Sustainable Development Goals (UN, 2021). The sustainability of the food production systems faces many challenges and as the Philadelphia farmer explained, there are multiple barriers and trade-offs small farms need to make to be included in the global value chain (Brandi, 2017). He and other small growers have found their niche with organic farming (see Fig. 6.1) but the dominant industrialised agricultural activities are subverting land use, causing soil degradation and biodiversity loss, polluting and exhausting the freshwater systems. Combined with the challenges posed by climatic extremes, the projected need for a 50% increase in production (FAO, 2017a) by 2050 for the growing global population (UN, 2019), animal health threats, such as foot-and-mouth disease, African swine fever, outbreaks of Newcastle disease in poultry, sheep and goat plague, the risk of superbugs and the propagation of emerging infectious zoonotic diseases associated with livestock – SARS (2002–2004), H5N1 (2008) and H1N1 (2009), the future of food security looks bleak. With food production becoming increasingly difficult and unpredictable in many parts of the world affected by droughts, floods, temperature extremes, heat and cold waves, dust storms, crop pest disease outbreaks and infestations, intensive precipitations, cyclones, typhoons, hurricanes, wildfires and adding to this the human impact of COVID-19, "[a]t no other point in history has agriculture been

Fig. 6.1 Organic market
in Pennsylvania,
USA. (Source: Authors)

faced with such an array of familiar and unfamiliar risks" (FAO, 2021, p. 33). Conventional agricultural approaches seeking increased yields are no longer a viable option.

The human response in defying situations like this is always to find technological solutions. Such technologies allow society to transition on a new trajectory that transforms the current practices for good. Most of these new technologies are less exploitative of the natural environment and use less land to produce the valuable nutrients required by humans. The focus on this chapter is on the technologies that are facilitating such a transition. We first briefly summarise why sustainability transitions in food production are needed and then reflect on the nature of these transformations in relation to the emerging 7th technological wave (Marinova et al., 2017). The actual new technological changes are organised in several groups, namely: vertical farming, artificial constructions or ponics, new approaches to agricultural practices (agroecology, agroforestry and the use of drones), nanotechnology in food production and application of information and communication technologies (ICTs), including artificial intelligence, 3D printing and blockchain. All this allows us to transition to clever food production and choices that are environmentally and nutritionally better.

The Need to Transition

The need to transition to clever food production is driven by an array of factors and we already covered many of them in earlier parts of the book. Here we particularly emphasise the restrictions associated with land availability on Earth to provide for nutritional food, resource exhaustion as a result from intensive farming and the need to link human well-being with the health of the planet.

Agricultural land currently uses more than one-third of the planet's terrestrial surface (37% of all land area, World Bank, 2018a) with only 11% being arable (World Bank, 2018b). Current trends in yield improvements will not be enough to satisfy the projected demand for food by 2050, even if the gaps between different agricultural production systems are closed (Bajželj et al., 2014). Expansion to new agricultural lands is undesirable as it would come at the price of further losses of biodiversity, soil degradation and increases in greenhouse gas emissions (Bajželj et al., 2014). Despite physical restrictions in the availability of land, substantial dietary changes toward meat and dairy products have occurred as a result from increased economic prosperity and urbanisation, and these food choices have even greater land and resource requirements (Keating et al., 2014). Dietary changes away from meat and dairy products towards nutrition-oriented agriculture are an essential part of the solution (Garnett, 2011; Keating et al., 2014). A study by Herrero et al. (2017) shows what the farmer from Philadelphia was explaining based on his personal observations – with the increase in farm size, the diversity of crops and nutrient production diminishes, especially that of the highly nutritious foods, such as vegetables and fruits, in favour of monocrops and livestock. This industrialisation of agriculture further challenges food security and the provision of adequate nutrients for the human diet.

Increasing demand for food inevitably leads to resource exhaustion. The basic natural resources of soil, water and air needed to sustain our food are threatened and degraded at a time when the burgeoning human population needs them most (Franzluebbers et al., 2020). Let's look at soils in particular as this natural resource is wrongly assumed to be renewable and therefore is often taken for granted. Plants obtain hydrogen, carbon and oxygen from the atmosphere, but they further need nitrogen, potassium and phosphorus from the soils and these are the primary nutrients allowing them to grow properly. However, humanity soon will be facing irreversible problems as these resources are being exhausted and exploited beyond repair. There is already evidence that we have exceeded the planetary boundaries for the use of *phosphorous* (Steffen et al., 2015) and the reserves of this finite resource can no longer sustain agriculture's exponentially growing demand. More than 10 years ago, the scientists' estimate was for phosphorous to be depleted within 50–100 years (Cordell et al., 2009, 2011) but agriculture is still progressively following the path of exhausting this non-renewable resource. Furthermore, anthropogenic addition of nitrogen, particularly in the form of fertilisers, is driving terrestrial ecosystems towards greater phosphorus limitations (Deng et al., 2017) which is additionally jeopardising future food security. The problem is worsening as at present industrialised farming is not adopting substitutes for phosphorus.

Nitrogen is the most abundant chemical element in the atmosphere and is also found in soils. Intensive farming uses added nitrogen in the form of fertilisers to increase yields. However, only a fraction of the nitrogen inputs is retained in the soils (40–60% in Australia, Agriculture Victoria, 2021) with the reminder released in the atmosphere where it combines with oxygen and forms nitrous oxide – a very potent and long-lived greenhouse gas. We have long exceeded the nitrogen planetary boundary and are now living in the zone of high risk (Steffen et al., 2015). *Potassium* is extracted from the soils by the plants, including their roots, and intensive farming does not allow for replenishment. Continuous harvesting gradually depletes the reservoirs of potassium in the soil. One 9-year study in Argentina shows that in addition to topsoils, subsoils are significantly affected by the long-term removal of this essential nutrient due to agricultural practices and some parts of the country had only 6 years left before the potassium budget is exhausted (Correndo et al., 2021). When it comes to the essential plant nutrients, it is no longer a matter of preventing further environmental deterioration and avoid causing harm; we need to find ways to heal the planet (Franzluebbers et al., 2020).

Often, particularly in the West and within the comfort of the urbanised environments, we do not realise how important and fragile the food system is. It is a major economic activity, accounting for 10% of the global GDP, representing $8 trillion in the world's supply chain and employing 1.5 billion people around the globe (Economist, 2020). It carries the task to secure food and feed everyone a high-quality diet as much as possible while minimising damage to our natural environment (Poore & Nemecek 2018; Willett et al., 2019), supporting planetary health (Marinova & Bogueva, 2019) and restoring ecosystems (Franzluebbers et al., 2020). With this complex agenda, there is no doubt that our food production must make a turn and reinvent itself in a completely new way using technological innovations and initiatives that disrupt the current destructive status quo and enhance the ecological, economic and social well-being responding to the planetary emergency. Right now, food production is at the crossroads of important sustainable development challenges and urgently needs to transition to a more sustainable, healthier, resilient and restorative future. Innovations and future technological applications are critical, can be transformational and can accelerate this transition (Herrero et al., 2020; McClements et al., 2021).

Transitioning

A sustainability transition is defined as a "radical transformation towards a sustainable society, as a response to a number of persistent problems confronting contemporary modern societies" (Grin et al., 2010, p. 1). The history of innovation can be understood as a process of linking together broader societal changes governed by an overarching socio-technological analytical framework that allows to successfully capture possible solutions to larger problems (Smith et al., 2005). Emerged in the 1930s, the idea of technological waves continues to describe the clustering of innovation activities around major technologies which revolutionise society and its production systems. Since industrialisation, six such waves have been described (Silva & Di Serio, 2016), related respectively to: Industrial Revolution, the Age of Steam, the Age of Electricity, the Age of Mass Production, the advent of Information and Communication Technologies and more recently, Sustainability (see also Fig. 6.2). Mostly concentrated in the developed countries, these waves have allowed enormous improvement in the quality of life of people and have been described as technological progress. However, they have also affected the foundations of global society allowing divisions and inequalities to persist with widening disparities (UNCTAD, 2021) and increased exploitation of the world's resources by the wealthy nations. The current planetary emergency is the result of such an unchecked technological progress.

We are now at the cusp of the Seventh Technological Wave which is already showing signs of emergence (Marinova et al., 2017). This wave is defined by the change of focus from exploitation of the planet and its resources to healing and regeneration (see Fig. 6.2). The COVID-19 pandemic precipitated advances in medical and human health-related technologies and the climate emergency is posing new challenges. This new wave is also a turning point in reconceptualising food and transitioning to clever technological solutions that are more sustainable and will allow to feed the growing human population while respecting and looking after the natural environment.

The current production of large amounts of inexpensive and standardised, often discretionary foods, is likely to remain in the past as we transition to sustainable food production. These future food systems need to use methods of production,

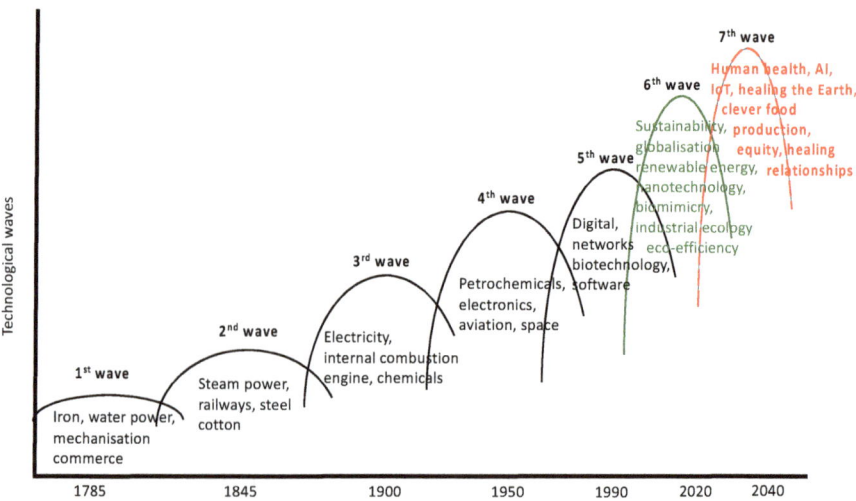

Fig. 6.2 Technological waves. (Note: *AI* artificial intelligence, *IoT* Internet of Things)

processes and systems that are truly sustainable, non-polluting, non-harming to the environment, conserving as much as possible non-renewable resources and energy in an economically efficient and safe for the producers and consumers way. They should not compromise the needs of the future generations but also should allow to deal with the current multifaceted dimensions of food inequalities that permeate countries, groups within society or individuals (UNCTAD, 2021).

According to McClements et al. (2021), there are numerous technological innovations the modern food industry applies to address some of the major challenges associated with generating an abundant, resilient, healthy and sustainable food supply and creating resilience of the food systems to pandemics, including COVID-19, and other disruptors. Technological solutions such as gene editing, biotechnology, nanotechnology, food architecture, robotics and automation, big data and machine learning, alternative proteins, sensor technologies, artificial intelligence (AI), green chemistry (McClements et al., 2021) as well as utilisation of polyculture permaculture systems, agroforestry and circular agriculture, growing without soil (e.g. hydroponics, aeroponics, hydroculture and nutrient film technique), vertical farming, the use of desalinated water and blockchain applications are all part of these technological innovations. These are all intended and have the potential to enhance the productivity and stability of human food supply while optimising agricultural land use, improving sustainability and efficiency throughout the supply chain.

Although the focus on food technologies is new within the context of technological waves, some of the ideas may have their roots in the past or simply mimic natural ecosystems. Let's look closer at them. Combinations between different approaches and technologies are also possible.

Vertical Farming

Vertical farming is based on a simple concept of the farm extending up rather than out (Al-Kodmany, 2018). This makes growing food an option suited to urban environments where the majority of human population now lives, using three types of vertical farming approaches, namely: construction of tall structures with growing beds at various levels and often relying on artificial lights; rooftop farming on commercial or residential buildings and dedicated multi-story farming building yet to be constructed anywhere in the world (Al-Kodmany, 2018). As this concept relies on creating conditions for growing plants that minimise land use, it does not rely on soil properties but on artificially constructed environments that optimise year-round production independent on weather or climate (Muller et al., 2017). An option in vertical farming is indoor farming where pesticide-free, soilless, nutrient-rich growing environments can be established (Despommier, 2013). Some of the world high-tech giants are among the pioneers in vertical farming described also as "skyscraper-" or "aerofarming" (Parkinson, 2016). Other existing structures, such as parking lots, abandoned warehouses, underground tunnels, undercover car parks and even shipping containers, can be used for vertical farming (Leblanc, 2020). Underground car parks in Paris, World War Two tunnels in London or dead shopping malls in USA (exacerbated by the COVID-19 pandemic) are examples of places in the urban environments where vertical farming can prosper (Park, 2020).

Soil is rarely used as a growing medium in vertical faming to avoid the spread of pests and diseases but also because it is too heavy and optimisation of nutrients is difficult. Techniques, such as aeroponics, aquaponics or hydroponics (see below), are commonly employed (Leblanc, 2020) with organic (peat moss, coconut husks, sphagnum moss or cellulose-based) and inorganic (clay pebbles, rocks, lava rocks or rockwool) growing material. These new technologies allow for food to be grown near consumers in high-tech climate-controlled environments where AI-powered containers can communicate and adjust growing conditions (Park, 2020).

The unusual growing spaces for vertical farming allow managing food demand using significantly less water and less space (Al-Kodmany, 2018). Organic and other crops can grow all-year-round with minimal cross-contamination and less exposure to chemicals and diseases due to continuous monitoring and testing with automation and sensors controlling quality and growth. Vertical farming often uses carefully controlled light, humidity, nutrient and temperature levels to optimise crop health and nutrient content (McClements et al., 2021). The opportunities vertical farming creates make the food system more resilient to weather conditions, calamities and other disruptions as well as shortens the supply chain to the consumers.

Despite its immediate benefits and promises to improve food security, natural processes, such as plant pollination, could be very difficult and costly with a heavy reliance on technology. Regardless of the controlled environment in which the food is grown, a short power loss or any other unforeseen technical issue could lead to devastating effects (Leblanc, 2020). There are also limitations related to production costs, especially electricity use for artificial lighting, control of the growing environment, including maintaining optimal temperature, humidity and carbon dioxide levels. For food produced in such a manner to be really sustainable, energy has to come from renewable sources, e.g. solar systems. Reliance on water needs to be optimised by capturing rainwater, using drip irrigation and recycling. Depreciation of the equipment and recycling of the materials used are also a consideration. Furthermore, not all plants are well-suited to be grown using vertical farming and making things even more complex, various species use different regions of the light spectrum (e.g. blue vs red or green) during the process of growth (Lawson & Kaya, 2018). This method of food production is generally more labour-intensive and this may be seen as a cost barrier; however, it is also an employment opportunity in the city.

The idea of vertical farming as eco-friendly and productive, allowing growing food quickly, locally and sustainably, is rapidly gaining momentum. However, it is not that new as similar concepts were applied in the past. The construction of the Hanging Gardens of Babylon built by King Nebuchadnezzar II (who reigned c. 605–c. 561 BCE) were considered one of the Seven Wonders of the Ancient World (Al-Kodmany, 2018). They represented a remarkable work of engineering art presented in an ascending series of multi-storey gardens, containing a vast variety of tree species, shrubs and vineyards, resembling a large green mountain built of mud bricks. Vertically layered growing techniques were also used for centuries and continue to be used in East Asia (Lawson & Kaya, 2018) for creating terraces for growing rice utilising mountainous land otherwise not suitable for farming. Their construction follows the natural contours of the mountains in a design that enables water to flow successively down from level to level decreasing soil erosion and surface runoff.

The term "vertical farming" as a type of controlled-environment agriculture was coined in 1915 by the geologist Gilbert Ellis Bailey who wrote a book with the same title in search of new methods for optimising cultivation (Bailey, 1915, p. 59). This book presents and supports the concept of vertical farming advocating for producing food year-round in stacked rows, close to the population centres, with no use of chemicals, while recycling materials and reducing water waste. Although the book gives limited information about the vertical farming technology itself, Bailey argued that farming hydroponically with constantly monitored and adjusted factors, such as temperature, nutrients, lighting, irrigation and air circulation, in a controlled vertical environment would provide economic and environmental benefits (Bailey, 1915).

In the 1980s, the Swedish ecological farmer and inventor of a spiral-shaped rail system for growing plants Åke Olsson proposed vertical farming as a way for growing vegetables in urban settings (Advantage Environment, 2009). In 1999, the modern concept of vertical farming was brought into action by the American microbiologist Dickson Despommier who described it as "mass cultivation of plant and animal life for commercial purposes in skyscrapers" (Despommier, 2010, p. 15). Nowadays vertical farming is particularly suited for urban environments where it allows to reconnect cities with agriculture and creates new patterns in food supply (Thomaier et al., 2015).

Is the future of farming indoors and in controlled environments? Many futurists believe this to be a viable option because of the ability to produce superfoods with the support of artificial intelligence, automation, robotics, optics and other precision technology (Lawson & Kaya, 2018). The future of farming is changing and vertical farming offers a pragmatic solution for cultivating local food in the world's rapidly growing cities (Parkinson, 2016). It takes away the uncertainty of agriculture by creating controllable growing environments and by doing so, it is becoming part of the food sustainability transition. The use of plastics in vertical farming, or any of the ponics methods, needs to adhere to circular economy principles of recycling and reuse in order to reduce pollution.

Ponics

Ponos (Πόνος) is the god of hard labour or toil in the Greek mythology (Definitions, n.d.-b) and the ending "-ponic/s" is used to describe an artificial structure or an ecosystem built with technology that does the work of feeding a plant (Definitions, n.d.-a). The meaning of the word "ponic" is to catch or take something from and the accompanying words or morphemes determine how the process works (Meaning

Directory, 2016). In relation to plants, ponics are growing systems that are not part of the lithosphere (the solid outer part) of the Earth (Meyer, 2017). Ponics are a range of controlled growing environments covering hydroponics (water), aeroponics (air), aquaponics (fresh water), aeroaquaponics (air and fresh water), maraponics (seawater), haloponics (brackish water on land), algaeponics (algae), digeponics (anaerobic) and vermiponics (worms) seen as technologies facilitating soilless food production. Although some of these systems boast many years of testing and exploration, research is still needed to expand the existing knowledge about their efficiency, ways to reduce current energy and infrastructure requirements, design aspects, capacity, capabilities and benefits.

Being methods for integrated, innovative food production, ponics technologies have also the pressing challenge of addressing regulatory issues. Harmonisation of regulations and legislation in relation to food safety and the export-import trade of produced products has commenced by the Food and Agriculture Organisation (FAO) in 2015, World Health Organisation (WHO) in 2017 and the European Union (EU) in 2016, but more work is needed (Goddek et al., 2019). The consumer certification system is also not unified. For example, in USA and Australia aquaponics products can be certified as organic, but the same status cannot be obtained in the EU (Goddek et al., 2019).

Let's explore some of the ponics methods further. They can form part of vertical farming or can stand on their own; in either case, they use much less land and soil than traditional agriculture. The roots of some of these artificial construction methods of growing food go back in history but recent technological progress is making them a viable option for sustainability transition.

Hydroponics

Hydroponics – or cultivating food in nutrients-enriched water, is an example of a method of growing plants with century-old traditions. The idea about gauging the importance of soil and water for plant growth has passed through many experiments starting from ancient Egyptian times (Torabi et al., 2012) and going through theoretical explanations, such as Francis Bacon's book *Sylva Sylvarum* on growing plants without soil published in 1627. In 1699, John Woodward published his experiments with producing spearmint, potatoes and other crops without soil and with water from springs and rivers (Time, 1937). In the eighteenth century, Julius von Schaps and Wilhelm Knop created standardised nutrient solutions adding 16 essential chemicals for plant growth, including Nitrogen (N), Potassium (K), Calcium (Ca) and Magnesium (Mg), to water (Panwar, 2020). Some of these solutions are still used today. The first to see its commercial potential and to coin the term hydroponics was Frederick Gericke from the University of California at Berkeley who claimed in the 1930s that this technique would revolutionise food production (Hydro Naturals, 2021).

Nowadays the cultivation of plants using nutrient-enriched water and the hydroponics method is providing advantages over traditional soil-based agriculture, including the elimination of insects, fungi, bacteria and other soil-related problems, it is less labour-intensive when managing larger areas of production and requires low maintenance (Al-Kodmany, 2018; Despommier, 2013). It is one of the most-favoured hi-tech production systems for expanding agricultural output in India, the world's second most populous country (Sankhalkar et al., 2019).

Aeroponics

A few years ago, a Tillandsia plant from Bromeliaceae family landed at home. The first reaction was to look for a spot to plant it in the garden, but luckily somebody pointed to us that this was an airplant. Plants like this can thrive in any environment, including rainforests, arid desserts, elevated mountainous areas, on rocks and in swamps and need no soil to survive. In fact, most Tillandsia species rely on their root systems to attach themselves to trees or rocks and absorb moisture and nutrients from the air through their leaves (Joyce, 2013). This was our introduction into the aeroponics world but the plant we had was not edible and was still using the natural environment as its habitat.

The basic principle in aeroponics is to grow plants in an enclosed environment without soil or any medium by spraying their dangling roots and lower part of the stem with an atomised nutrient-rich solution. Frits Warmolt Went, a Dutch biologist who coined the word "aeroponics", grew tomatoes and coffee plants with their roots suspended in the air using a nutrient mist (Stoner, 1983). Aeroponics has allowed researchers to study the plants' root morphology, nutrient uptake, reaction to drought and flood stress as well as responses to variations in oxygen and/or carbon dioxide concentrations (Clawson et al., 2000).

Until recently, the potential of this method was seen as attractive mainly for growing food in space. Created and patented by Richard J. Stoner II in the late 1980s (NASA Spinoff, 2006), the aeroponics process of growing hanging in the air plants seemed like an extraordinary uninterrupted, planted and harvested all-year-round way for clean, efficient and rapid food production (Kotzen et al., 2019; Lakhiar et al., 2018). All plants produced with aeroponics absorb more minerals and essential elements, making them healthier and potentially more nutritious. The developed in the 1990s high performance aeroponic system ordered by the US National Aeronautics and Space Administration (NASA) proved to be successful both in space and on Earth. It was used to grow plants in space on the international Mir space station (Pure Greens Arizona, 2020). According to NASA, in aeroponic growing systems plants grow up to three times faster and with more consistent yield compared to these produced in soil (Kotzen et al., 2019). There is a lot of enthusiasm surrounding this method with NASA believing that with "technology like this, jumping over the moon (or other outer spaces) won't be reserved for fairy tales" (NASA, 2007, para. 14).

Growing in the air without the need for soil or any substrate culture using aeroponic systems offers roots room to breathe and tailored care can be applied, specified and based on the individual plant's requirements (Lakhiar et al., 2018). Compared to traditional agriculture, the technology allows growers employing the aeroponics method to reduce water usage by 98%, fertiliser application by 60% without any need for pesticides and while maximising crop yields by 45–75% (NASA Spinoff, 2006). According to Lakhiar et al. (2018, p. 338), aeroponics "is considered the best plant growing method for food security and sustainable development". It is seen as the most efficient, economical and convenient plant-growing technique compared to other soil-based and soilless methods (Lakhiar et al., 2018) offering a lot of promise for sustainable transition in food production.

Aquaponics

Aquaponics, a term coined in the 1970s, is another more sustainable farming method for efficient food production. It is perceived as an emerging food production technology with minimal impact on the natural environment (Kasozi et al., 2021) presenting a way that can aid addressing the planetary boundaries and the Sustainable Development Goals, especially as it allows for the recycling of nutrient resources and waste (Goddek & Korner, 2019). Aquaponics combines fish and soilless production of edible plants in a single loop (coupled) or two/multiple-loop (decoupled) system (Goddeck et al., 2019). Decoupled aquaponics systems (which use water from the fish for the plants but not vice-versa) separate the growth of fish from plants as they require different types of nutrients and tend to put more emphasis on obtaining animal protein.

Fish are being fed while the dissolved nutrients from uneaten fish feed and faeces are broken down into organic matter utilising microbes to convert them into nitrogen and phosphorous into bioavailable forms for use by plants (Joyce et al., 2019). The edible plants remove nutrients from the water minimising the need for replacement and there are reduced requirements to use fertilisers. Aquaponics technology is very appropriate for urban areas as it can concentrate food production in unusual spaces, such as roofs, schools, development sites or other places that normally will not be used for the purpose of growing foodstuff (van Gorcum et al., 2019). It can also be developed as a social enterprise.

As with many of the technologies in this group, some of the ideas can be traced back in time. The Aztec Indians around 1100 BC created their special floating gardens. These artificial agricultural islands were called "chinampas" and they provided a natural combination of intertwined plant roots, reeds with stakes beneath the lake's surface and sediment from lake-bottoms forming a natural habitat of nutrients for the crops with optimal moisture retention (Panwar, 2020). The "chinampas" system is considered the first form of aquaponics for agricultural use (Espinal & Matulic, 2019). Around 5 CE, aquaponics was also known to the Chinese farmers, especially from Yunnan, who were cultivating rice in paddies, flooded fields, together with fish (Espinal & Matulic, 2019). Similar polyculture farming techniques were integrated in many Far Eastern countries (East and Southeast Asia) to grow fish. Their reinvention now is aimed at improving the sustainability and quality of food production.

Despite a lot of promising potential, more research is needed to understand the complex microbiome of aquaponics, including biofiltration microorganisms, bacterial composition and diversity, biofilter establishment and cycling, and most importantly, avoiding human pathogens to guarantee the safety of aquaponics products (Kasozi et al., 2021). Aquaponics systems are also energy-intensive and the use of renewable power is essential to improve their environmental footprint. They also require special infrastructure and feed for the fish. Further potential challenges may emerge in controlling diseases, the spread of weeds, fungal and bacterial/algal contaminants (Joyce et al., 2019). Application of antibiotics and pesticides to maintain the health of fish and plants is also common.

Vertical aquaponics are also being designed (Kotzen et al., 2019). Combining aquaponics and aeroponics in *aquaeroponics* is a possibility that some are keen to explore. More research is needed in this area for such a food technology to become viable. At the moment, there are technical problems to be resolved, such as clogging of the mist sprayers and the required fine filtration (Espinal & Matulic, 2019). The development of pathogens in an airy wet environment is another serious concern (Espinal & Matulic, 2019).

Algae and Algaeponics

Algae are known as one of the oldest plants in the world, first appearing 3.5 billion years ago and consumed by humans for thousands of years (Koyande et al., 2019). Both, macroalgae (referred to as seaweed) and microalgae (single-cell organisms), are nonflowering plants with no stems, roots or leaves which grow in water. Found in most environments but predominantly aquatic, microalgae are producing 50% of the oxygen in the world and capturing substantial amounts of CO_2 with up to 50% of their biomass being carbon (Kotzen et al., 2019). They are also one of the most promising sustainable sources of food ingredients, rich in numerous valuable health-beneficial compounds, such as carbohydrates, polyunsaturated fatty acids, essential minerals and vitamins (Van Loey et al., 2020). Algae contain similar amounts of proteins compared to animal-based sources, such meat, milk and eggs, or plant-based foods, such as soya, wheat and legumes, but generate much higher yields. A study by Wageningen University reports yields from microalgae at 4–15 tonnes/ha/year and from macroalgae at 2.5–7.5 tonnes/ha/year compared to 1.1 tonnes/ha/year for wheat, 0.4–0.5 tonnes/ha/year for quinoa, 0.4–0.75 tonnes/ha/year for oat and 1–2 tonnes/ha/year for legumes (van Krimpen et al., 2013). Being exposed to free radicals and high oxidative stress, algae develop antioxidants and pigments, such as chlorophylls, carotenes and phycobiliproteins, which are beneficial for humans (Koyande et al., 2019).

Algae have been consumed as food and medicine in China for millennia. After the first microalgal cultivation of Chlorella vulgaris in 1890 (Borowitzka, 1999), many years have passed but algae are still an underexploited food source for human consumption. In Japan microalgae have been commercially available as healthy foods since the 1950s (Sathasivam et al., 2019). Food supplements containing algae have gained popularity in Mexico, Taiwan, Israel and the Western world, but the potential of this nutritional source is still underused.

An advantage of marine algae is that they do not need land or freshwater while microalgae can be grown on non-agricultural land. As algae have a very high photosynthetic efficiency, they grow fast and build substantial biomass that is resistant to various contaminants (Dębowski et al., 2020). Microalgae can also grow under limited nutrient conditions and have the ability to adapt to a wider range of environments (Gordon & Polle, 2007). In addition to being a good nutritional source, algae can have other industrial uses in bioenergy production, treatment of sewage and leachate, neutralisation of sludge and waste as well as animal feed.

Algaeponics are constructed systems for growing algae as food commercially. In that sense they differ from the other ponics which aim to grow familiar edible plants using soilless methods. Different tanks, ponds, large bags, cascade systems, fermenters and closed photobioreactors are used to commercially culture a wide spectrum of algae. They can also be grown as biofilms. It is important to control for light, temperature and critical nutrients, such as nitrogen (N), phosphorous (P) and carbon (C), as well as other elements that affect microalgal growth and metabolism (Kotzen et al., 2019). Although technologically algaeponics can be seen as a subset of aquaponics, the high potential algae hold as a novel food deserves it to be distinguished as a separate option in the sustainability transition.

Algae can be beneficial for improving water quality in any aquaponics systems, to control the pH, generate dissolved oxygen, produce polyunsaturated fatty acids as a value-added fish feed and build resilience in the system (Addy et al., 2017). The abundance of proteins and other essential nutrients in microalgae can be the basis for a massive algae-based food industry, dedicated towards commercialisation of healthy and functional foods.

Vermicomposting and Vermiponics

Vermicomposting is the use of worms (vermiculture) in converting organic waste into fertiliser while vermiponics refers to the use of worms in hydroponic systems. Protected by law during Cleopatra's reign, earthworms are still one of the most valuable and useful creatures looking after soil nutrients and delivering optimum plant health (Goddek et al., 2019). Earthworms are praised in agriculture and horticulture as they maintain soil health transporting nutrients and minerals from below to the surface via their waste and creating tunnels that aerate the ground (Kotzen et al., 2019).

In modern composting with worms, developed by Mary Appelhof in the 1970s and 1980s, organic matter of different size is processed, digested and converted into a rich liquid fertiliser (Kotzen et al., 2019). The final product of vermicomposting is regularly and successfully used in many gardens in Australia and worldwide. In vermiponics, which is a relatively new technique, red wriggler worms (also known as tiger worms) are introduced in a hydroponic system to break down any solid waste and detritus from the plants providing nutrients.

Both, vermicomposting and vermiponics are not producing directly food but facilitating the process of growing plants. They belong to the class of the emerging most-needed technologies for better and clever food production that respond to the current sustainability challenges. The environment and climate emergency however also calls for revision and reconceptualising of the current agricultural practices to make them more supportive of planetary health.

Better Agricultural Practices

The Philadelphia farmer about whom we spoke earlier is operating an organic farm trying to find the right balance between what he takes from and gives to nature. It is not an easy task when "essential natural resources… are being undermined by agricultural practices that continue to deplete the soil resource base, pollute freshwater and coastal estuaries needed for life support, reduce habitat to support biodiversity, and emit harmful greenhouse gases that compromise our ability to withstand changes to the climate" (Franzluebbers et al., 2020, p. 1). There are however some possible solutions and general principles which if adopted would help agriculture transitioning towards being more sustainable. Let's look at some of them, namely agroecology, agroforestry and regenerative agriculture.

Agroecology

Agroecology links agricultural production to the ecological processes and "includes all the techniques that allow agricultural practices to be more respectful of the environment and its ecological specificities" (youmatter, 2020, para. 2). The term appeared in 1928 with the work of the American agronomist Basil Bensin and the concept was progressively developed through the second half of the twentieth century (youmatter, 2020) culminating with its endorsement by FAO in 2018 (FAO, 2018) as a transition to sustainable food and agricultural systems and in response to the global 2030 Sustainable Development Goals (SDGs). Agroecology now covers principles and approaches that guide human beings to live in harmony with nature and not against it (Franzluebbers et al., 2020).

The agroecology approach should always be localised with different implementation at the level of the individual farm or globally. For example, establishing ecological corridors to facilitate links between habitats could be interpreted differently at a local and global level. Livelihoods are also important with social as well as ecological dimensions now included in the ten specific agroecology elements endorsed by FAO (2018):

- diversity – increased spatial diversity, rotation of crops and complementary species in order to maintain ecosystem services, including soil health and pollination;
- co-creating and sharing of knowledge – knowledge needs to be context-specific blending indigenous and traditional wisdom, global scientific progress and practical learnings;
- synergies – paying attention to co-designing systems where species complement each other, such as nitrogen fixation from legumes through crop rotation;
- efficiency – striving to produce more nutritious food with less external resources enhancing existing biological processes;
- recycling – recycling of biomass, nutrients and water within the production systems reducing pollution and minimising waste;
- resilience – enhancing the resilience of people, communities and ecosystems through diversification and the ability to recover after from disturbances;
- human and social values – improving livelihoods, equity and social well-being creating meaningful opportunities for all, including the youth;
- culture and food traditions – achieving food security and supporting cultural identities through nutritious, healthy, diversified and culturally appropriate diets;
- responsible governance – governance should support the transition to sustainable food and agricultural systems through transparent, accountable, inclusive, responsible and efficient mechanisms at the locally appropriate level;
- circular and solidarity economy – reconnecting producers and consumers within the planetary boundaries.

We should be seeking healing of the planet and we have the capacity to help by changing our practices. Nature can recover and rejuvenate if given a chance as we witnessed this in some places around the world during the COVID-19 lockdowns. At the end of 2020, there was 500% decrease in sewage and industrial effluents in rivers, the levels of dissolved oxygen, biological oxygen demand and pH of river water improved by 79%, 30% and to 7.9 respectively and noise level was reduced 35–68% all over the world (Arora et al., 2020). Most importantly, wild life on the land, rivers and oceans had the opportunity to also recover. Rather than needing a pandemic to mobilise human activities, we could change our agricultural practices and dietary preferences to give nature a chance. What some describe to be a diet, health and environment trilemma (Clark et al., 2018), essentially should form part of one unified sustainability agenda that allows the Earth to heal.

The three targets of the post-2020 agroecology agenda are (Wanger et al., 2020):

- Target 1 – reducing the threats to biodiversity;
- Target 2 – meeting people's needs through sustainable use and benefit sharing; and

- Target 3 – applying tools and solutions for implementation and mainstreaming of agroecology, including zero deforestation and a global transition to clever food production systems.

Agroforestry

A symbiosis between natural and created land-use management systems is the foundation of agroforestry which is defined simply as the interaction between agriculture and trees (World Agroforestry, 2021). In an ecologically beneficial way, agroforestry aims to balance protection of the natural environment with the production of nutritious food and the use of trees for fruits, nuts, timber and other commercial purposes. It is a model that rejects the clearing of native vegetation for the purpose of establishing agricultural land or pastures and instead integrates food production within wooded areas while preserving the multiple benefits of trees in sequestering carbon, mitigating climate changes, maintaining the water table, protecting against land erosion, allowing different levels of biodiversity, building soil organic matter, providing food, habitat and shelter for a variety of species as well as satisfying aesthetic and recreational requirements, to mention a few. Crops are grown around and in between trees and shrubs, increasing the biodiversity of agricultural landscapes. In a similar ecological beneficial way, agroforestry contributes to biodiversity conservation and water quality improvement as well as provides for cultural and spiritual needs (Lovell et al., 2017).

Agroforestry allows to mix crop and livestock systems, however, the scale of animal husbandry needs to be significantly reduced by at least 50% (Tirado et al., 2018; Willett et al., 2019). The approach is suited for forests located in a variety of geographical areas – from the tropics to temperate areas. Nitrogen-fixing plants can help maintain soil fertility and in addition to native species, the trees can include fruit- and nut-bearing varieties. In some parts of the world, agroforestry is combined with tea or coffee bushes and medicinal plants. While agroforestry is built mainly on scientific principles to achieve environmental and economic benefits, *permaculture* (originating from permanent agriculture) adopts a similar ecological approach extending it to creating nurturing relationships between people and their natural environment (Weiseman et al., 2014).

Regenerative Agriculture

Regenerative agriculture aims to restore the health of the soil and the entire ecosystem by improving the natural resources rather than depleting them (Climate Reality Project, 2019). Restoring the natural environment contributes to carbon sequestration, reduces water use and the application of fertilisers. Key agricultural techniques include conservation or no tillage, maintaining diversity of crops, using rotation and cover crops for nitrogen fixation and minimising physical disturbance, e.g. through the application of fertilisers. Farmers should be at the forefront of protecting and restoring the soil fertility and the Rodale Institute (2014) estimates that more than 100% of all carbon dioxide emissions in the Earth's atmosphere can be sequestered with available and inexpensive organic management practices reversing the greenhouse effect. Although regenerative organic agricultural systems have generally lower yields than conventional farming, they perform better economically and provide significant environmental benefits by maintaining planetary health (Durham & Mizik, 2021). This indicates a consumer-driven transition towards better food sustainability.

Although some argue that the role of livestock in regenerative agriculture should be to imitate the behaviour of wild animal herds (Massy, 2017), it is clear that the current large numbers cannot be sustained without compromising the ability of the soil and land to rejuvenate. Furthermore, the livestock species raised in a place, such as Australia, particularly cattle, are very different from the native marsupial animals (kangaroos, wombats, wallabies, bandicoots, Tasmanian devils or koalas), monotremes (echidnas and platypuses) or the dingos (which arrived to the Australian continent more than 8000 years ago). The digestive systems or ruminant animals emit large quantities of methane and livestock is ill-suited to withstand the weather calamities associated with climate change. Other more suitable approaches are needed to respond to the current planetary emergency.

Nanotechnology and Food

Nanotechnology, one of the leading technologies of the 6th technological wave (see Fig. 6.2), is also finding its way to food. Green methods of synthesising nanoparticles using plant extracts are perceived as being eco-friendly, simple, convenient and contributing towards improving crop growth and reducing pollution (Ghidan & Al Antary, 2019). Nanosensors and nanodevices can assist with precision farming techniques that reduce the use of fertilisers and pesticides as well as optimise water requirements. There are however concerns about the safety of nanomaterials with more toxicology studies needed about their impacts on the soil, water, natural environment, other species and people, particularly as this is a relatively new field compared to chemical fertilisers and drugs (Sekhon, 2014).

Nano-agrichemicals

The large surface area of the nanoparticles allows them to address some of the challenges unmet by other chemical or physical methods of fertilising or providing pest control (Ghidan & Al Antary, 2019). They can be used for controlled and targeted release of different chemicals for improved plant growth, as pesticides, fungicides or insect repellents. The substances made up from nanoparticles can be absorbed easily by the plants and be programmed for release at a specific time (Kah, 2015). Such products are described as nano-agrochemicals as they combine nanotechnology with agrochemicals as eco-friendly alternatives to conventional chemical fertilisers, pesticides, fungicides, insecticides and herbicides (Qazi & Dar, 2020). There are high expectations that nanotechnology will improve agricultural practices by also using intelligent nanotools (e.g. nanosensors), allowing for precise control management of pesticides, fertilisers and water, improving the quality and quantity of production yields and overall reducing the impact of modern agriculture on the environment (Kah, 2015), representing a revolution in food cultivation. Big companies, such as Monsanto, Syngenta and BASF invested in the past in this field (Kah, 2015).

Agro-nanobiotechnology is another new term which describes the potential of nanotechnology to control biological processes, preserve environmental health and facilitate agricultural production (Fernández-Luqueño et al., 2018). This includes monitoring, remediation of contaminated sites and restoring land for agricultural use. Again, these new technologies require appropriate trials and legislation to guarantee they are safe and can contribute towards a sustainability transition. Reducing concerns about fate, transport, bioavailability and toxicity of nanoparticles can open the road of acceptance of these technologies and their adoption in agriculture (Mishra et al., 2017).

Nanofoods

When nanotechnologies or nanoparticles are used in the process of cultivation, production, processing, preservation, quality control or packaging, food becomes nanofood (Sekhon, 2014). Technological advances have made it possible to encapsulate bioactive compounds at the nanoscale (Garba & Ismail, 2020). Such nano-size additives can be added to improve the taste, colour and texture of food but also to make it more nutritional by carrying vitamins, minerals and phytochemicals or to prevent spoiling, increase durability and extend shelf life of products.

These promising applications still need proper risk assessment as the processes that are currently in place are not specifically designed for agri-food nanomaterials. Furthermore, harmonised international regulations governing organisations and individuals working with nanomaterials and nanoparticles in food need to be put in place to protect planetary health (Meghani et al., 2020).

Computers and Food

The tremendous progress made in the field of information and communication technologies (ICT) combined with constantly improving mechanics, robotics, optics and automation – technologies from the 5th technological wave (see Fig. 6.2) are similarly rediscovering food- and agriculture-related applications. It is impossible to cover all developments as they are gradually permeating and reshaping traditional agriculture to reduce its environmental impacts and achieve security of food with high nutritional qualities. Below are some of the technologies which help the transition to clever food production.

3D Bioprinting

Given that animal-based meat has the highest environmental footprint, 3D bioprinting is the start of a new era offering an alternative approach. Originally developed for nonbiologic applications in the 1980s by Charles Hull, the founder of 3D Systems in California, USA (Hoffman, 2020), the method was intended to simply model potential end products made from different materials under more traditional techniques. Starting as organ printing aimed at medical applications, today 3D bioprinting covers bioinks and a range of technologies, including extrusion, inkjet, laser-assisted and stereolithography bioprinting (Choudhury et al., 2018) which can help replace animal-based foods, such as meat. Green bioprinting involves plant cells and can be used for designer plant-based foods. It adds another dimension to the sustainability transition in food production offering a new manufacturing platform for the fabrication of *in vitro* tissue (Choudhury et al., 2018). From US$ 263.8 million investment in 2015, amid the COVID-19 crisis of 2020 the total bioprinting market reached US$768.9 million (Business Wire, 2021) (some estimate it at US1.4 billion, Grand View Research, 2021) and more than a quarter of the applications are related to food, including meat but also other products associated with high greenhouse gas emissions, such as cheese and chocolate. Bioprinting is expected to increase to US$2.8 billion by 2027 with USA and China leading the progress and players, such as Japan, Canada and Germany contributing to the geographic spread of these technologies (Business Wire, 2021).

Currently, 3D bioprinting is capable to create fully functioning cell-based complex organs, generate custom-made bone, cartilage, blood vascular network and tissues (Zhang & Wang, 2017; Choudhury et al., 2018). It offers the opportunity to design novel food products with improved nutritional value and sensorial profile (Dick et al., 2019). Many companies are pursuing to reconstruct the best 3D bioprinted meat, such as the Israeli-based start-ups Redefine Meat and Aleph Farms Ltd. which are experimenting with 3D printing to recreate the texture, muscle, blood and fat of a real steak (Ramachandraiah, 2021; Bandoim, 2021; Meisenzahl, 2020) although challenges exist. Obstacles to acceptance of these foods are not just the unexplored consumer perceptions but also that there are no existing regulations or regulatory frameworks in place to support these innovations (Bandoim, 2021).

Drones, Satellites, AI and Others

There is a vast range of new ICT-based technologies which are transforming the way food is produced. For example, drones equipped with high-tech cameras improve the monitoring of plants to identify nutritional deficiencies, the need for irrigation, presence of pests and diseases. They are also used to plant seeds, to count trees and farmers are increasingly embracing this new technology (Sheehan, 2019). Satellites are used for better forecasting of weather as well as for monitoring crop health and water management. Driverless autonomous tractors and other machinery are also on the horizon to improve farming efficiency. Artificial intelligence (AI) can facilitate many complex and routine jobs farmers have to do and when combined with robots and automation, can actually perform a range of tasks associated with food production while providing optimal solutions, cost savings, assessing and managing risk (Lenniy, 2021). Foods can be designed using AI in a way that improves their healthiness by reducing high levels of salt, sugar and fats, increasing the content of vitamins, minerals, prebiotics and probiotics (McClements, 2020), and eliminating concerns about over-processing (Fardet & Rock, 2020). It is also possible to design foods that are digested slower or release their ingredients at particular locations along the human gastrointestinal track reducing appetite (McClements, 2020). The Internet of Things (IoT) with sensors and geographic information systems (GIS) further expands the possibilities not only for conventional but also for vertical farming. As part of the 7th technological wave, we are likely to see a large clustering of new technologies and applications related to food production as well as for rejuvenating the health of the planet.

Blockchain

Although blockchain technology does not change the way food is produced, it adds transparency for consumers and downstream users about the production process. Representing a distributed database – a ledger of accounts and transactions created by the participants, it stores the information in blocks chained together. The technology allows traceability as well as on-the-spot financial transactions. It has various potential applications in agriculture, particularly in securing food supply chains, food traceability and safety, proving sustainability and high-value practices and increasing payment security for farmers.

The different blockchain applications present a platform for a reliable source of otherwise costly information revealing the exact state of any participating farms, inventories, food provenance and contracts in agriculture without using an intermediary or a middleman/middlewoman in the agriculture sector (Xiong et al., 2020). It facilitates the tracking of products along the agri-food supply chain and is expected to become an "efficient and robust mechanism for enhancing food trace-ability and a transparent and reliable way to validate quality, safety, and sustainabil-ity, of agri-foods" (Xu et al., 2020, p. 153). By doing this, blockchain facilitates better agri-food management, including quality, safety and environmental perfor-mance, but there are risks with data security and data integration. It should also be acknowledged that computers are energy-intensive and if renewable sources are not used, this contributes to the environmental footprint of food.

Concluding Remarks

The unprecedented and rapid advances in food production are transforming what we eat and facilitating a sustainability transition (McClements, 2019). Towards the end of 2020, we had the opportunity to engage with Professor David Julian McClements from the Department of Food Science at the University of Massachusetts. He spoke about food architecture, gene editing, nanotechnology and artificial intelligence as technologies used to address the modern food challenges of feeding the growing global population, reducing greenhouse gas emissions, elimi-nating waste and achieving a transition to sustainable food production. There was a question from the audience whether the nano-enhanced food products should be considered as "Frankenfoods" or as a revolutionising new technology. People from the audience and elsewhere are concerned, curious and excited about these and the other technological innovations. The answer to this question lays in the future. What is certain now is that we can no longer continue with the current trends in food sys-tems as they have indeed gradually turned into Frankenstein solutions to the basic problem of feeding the global population.

Similar to Frankenstein who stopped noticing the seasons absorbed in his inven-tion, the food and agriculture industry deeply engrossed in the ambition to produce more foodstuff and generate profits, no longer sees the impacts of its actions on the natural environment and people, or chooses not to see or admit them. Governments and societies allowed this to happen until this point in time when we are faced with a planetary emergency. The evolving food technologies offer promising new solu-tions. However, it is up to us as consumers and producers whether the new clever food technologies would become Frankenstein inventions or allow a transition on a better pathway. We hope for the latter.

References

Addy, M. M., Kabir, F., Zhang, R., Lu, Q., Deng, X., Current, D., Griffith, R., Ma, Y., Zhou, W., Chen, P., & Ruan, R. (2017). Co-cultivation of microalgae in aquaponic systems. *Bioresource Technology, 245*(Pt A), 27–34. https://doi.org/10.1016/j.biortech.2017.08.151

Advantage Environment. (2009). *Vertical greenhouse for urban agriculture.* http://advantage-environment.com/food/vertical-greenhouse-for-urban-agriculture/

Agriculture Victoria. (2021). *Nitrogen fertilisers – Improving efficiency and saving money.* https://agriculture.vic.gov.au/climate-and-weather/understanding-carbon-and-emissions/nitrogen-fertilisers-improving-efficiency-and-saving-money

Al-Kodmany, K. (2018). The vertical farm: A review of developments and implications for the vertical city. *Buildings, 8*(2), 24. https://doi.org/10.3390/buildings8020024

Arora, S., Bhaukhandi, K. D., & Mishra, P. K. (2020). Coronavirus lockdown helped the environment to bounce back. *Science of the Total Environment, 742*, 140573. https://doi.org/10.1016/j.scitotenv.2020.140573

Bailey, G. E. (1915). *Vertical farming.* E. I. du Pont de Nemours Powder Co. https://archive.org/details/cu31924000349328

Bajželj, B., Richards, K. S., Allwood, J. M., Smith, P., Dennis, J. S., Curmi, E., & Gilligan, C. A. (2014). Importance of food-demand management for climate mitigation. *Nature Climate Change, 4*(10), 924–929. https://doi.org/10.1038/nclimate2353

Bandoim, L. (2021). *World's first 3D bioprinted and cultivated ribeye steak is revealed.* Forbes. Available at: https://www.forbes.com/sites/lanabandoim/2021/02/12/worlds-first-3d-bioprinted-and-cultivated-ribeye-steak-is-revealed/?sh=3c3be7424781

Borowitzka, M. A. (1999). Commercial production of microalgae: Ponds, tanks, tubes and fermenters. *Journal of Biotechnology, 70*(1–3), 313–321. https://doi.org/10.1016/S0168-1656(99)00083-8

Brandi, C. A. (2017). Sustainability standards and sustainable development – Synergies and trade-offs of transnational governance. *Sustainable Development, 25*, 25–34. https://doi.org/10.1002/sd.1639

BusinessWire. (2021). *Global 3D bioprinting market trajectory & analytics report 2021 – research-andmarkets.com.* https://www.businesswire.com/news/home/20210628005296/en/Global-3D-Bioprinting-Market-Trajectory-Analytics-Report-2021%2D%2D-ResearchAndMarkets.com

Choudhury, D., Anand, S., & Naing, M. W. (2018). The arrival of commercial bioprinters – Towards 3D bioprinting revolution! *International Journal of Bioprinting, 4*(2), 139. https://doi.org/10.18063/IJB.v4i2.139

Clark, M., Hill, J., & Tilman, D. (2018). The diet, health, and environment trilemma. *Annual Review of Environment and Resources, 43*, 109–134. https://doi.org/10.1146/annurev-environ-102017-025957

Clawson, J. M., Hoehn, A., Stodieck, L. S., Todd, P., & Stoner, R. J. (2000). *Aeroponics for space-flight plant growth.* Aeroponics Daily. https://aeroponicsdiy.com/nasa-review-of-aeroponics/

Climate Reality Project. (2019). *What is regenerative agriculture?* https://www.climaterealityproject.org/blog/what-regenerative-agriculture

Cordell, D., Drangert, J.-O., & White, S. (2009). The story of phosphorus: Global food security and food for thought. *Global Environmental Change, 19*(2), 292–305. https://doi.org/10.1016/j.gloenvcha.2008.10.009

Cordell, D., Rosemarin, A., Schroder, J., & Smit, A. (2011). Towards global phosphorus security: A systems framework for phosphorus recovery and reuse options. *Chemosphere, 84*(6), 747–758. https://doi.org/10.1016/j.chemosphere.2011.02.032

Correndo, A. A., Rubio, G., García, F. O., & Ciampitti, I. A. (2021). Subsoil-potassium depletion accounts for the nutrient budget in high-potassium agricultural soils. *Scientific Reports, 11*, 11597. https://doi.org/10.1038/s41598-021-90297-1

Dębowski, M., Zieliński, M., Kazimierowicz, J., Kujawska, N., & Talbierz, S. (2020). Microalgae cultivation technologies as an opportunity for bioenergetic system development – Advantages and limitations. *Sustainability, 12*, 9980. https://doi.org/10.3390/su12239980

Definitions. (n.d.-a). https://www.definitions.net/definition/ponics

Definitions. (n.d.-b). *Ponos*. https://www.definitions.net/definition/Ponos

Deng, Q., Hui, D., Dennis, S., & Reddy, K. C. (2017). Responses of terrestrial ecosystem phosphorus cycling to nitrogen addition: A meta-analysis. *Global Ecology and Biogeography, 26*, 713–728. https://doi.org/10.1111/geb.12576

Despommier, D. (2010). *The vertical farm: Feeding the world in the 21st century*. Thomas Dunne Books.

Despommier, D. (2013). Farming up the city: The rise of urban vertical farms. *Trends in Biotechnology, 31*(7), 388–389. https://doi.org/10.1016/j.tibtech.2013.03.008

Dick, A., Bhandari, B., & Prakash, S. (2019). 3D printing of meat. *Meat Science, 153*, 35–44. https://doi.org/10.1016/j.meatsci.2019.03.005

Durham, T. C., & Mizik, T. (2021). Comparative economics of conventional, organic, and alternative agricultural production systems. *Economies, 9*, 64. https://doi.org/10.3390/economies9020064

Economist. (2020). *How to feed the planet: The global food supply chain is passing a severe test*. The Economist. https://www.economist.com/leaders/2020/05/09/the-global-food-supply-chain-is-passing-a-severe-test

Espinal, C. A., & Matulic, D. (2019). Recirculating aquaculture technologies. In S. Goddek, A. Joyce, B. Kotzen, & G. M. Burnell (Eds.), *Aquaponics food production systems: Combined aquaculture and hydroponic production technologies for the future* (pp. 35–76). Springer. https://link.springer.com/content/pdf/10.1007%2F978-3-030-15943-6.pdf

Fardet, A., & Rock, E. (2020). Ultra-processed foods and food system sustainability: What are the links? *Sustainability, 12*, 6280. https://doi.org/10.3390/su12156280

Fernández-Luqueño, F., Medina-Pérez, G., López-Valdez, F., Gutiérrez-Ramírez, R., Campos-Montiel, R., Vázquez Núñez, E., Loera-Serna, S., Almaraz-Buendía, I., Razo, D., Oscar, & Madariaga-Navarrete, A. (2018). Use of agronanobiotechnology in the agro-food industry to preserve environmental health and improve the welfare of farmers: Modern agriculture for a sustainable future. In *Agricultural nanobiotechnology*. https://doi.org/10.1007/978-3-319-96719-6_1

Food and Agriculture Organisation of the United Nations (FAO). (2017a). *The state of food and agriculture 2017: Leveraging food systems for inclusive rural transformation*. http://www.fao.org/policy-support/tools-and-publications/resources-details/en/c/1046886/

Food and Agriculture Organisation of the United Nations (FAO). (2017b). *UN climate change conference acknowledges agriculture sectors' critical role*. http://www.fao.org/news/story/en/item/1068313/icode/

Food and Agriculture Organisation of the United Nations (FAO). (2018). *The 10 elements of agroecology: Guiding the transition to sustainable food and agricultural systems*. http://www.fao.org/3/I9037EN/i9037en.pdf

Food and Agriculture Organisation of the United Nations (FAO). (2021). *2021: The impact of disasters and crises on agriculture and food security*. http://www.fao.org/3/cb3673en/cb3673en.pdf

Franzluebbers, A. J., Wendroth, O., Creamer, N. G., & Feng, G. G. (2020). Focusing the future of farming on agroecology. *Agricultural and Environmental Letters, 5*(1), e20034. https://doi.org/10.1002/ael2.20034

Garba, U., & Ismail, B. B. (2020). Supercritical fluid techniques to fabricate efficient nanoencapsulated food-grade materials. In U. Hebbar, S. Ranjan, N. Dasgupta, & R. Kumar Mishra (Eds.), *Nano-food engineering* (pp. 25–47). Springer. https://doi.org/10.1007/978-3-030-44552-2_2

Garnett, T. (2011). Where are the best opportunities for reducing greenhouse gas emissions in the food system (including the food chain)? *Food Policy, 36*, S23–S32. https://doi.org/10.1016/J.FOODPOL.2010.10.010

Ghidan, A. Y, & Al Antary, T. M. (2019). *Applications of nanotechnology in agriculture*. IntechOpen. https://www.intechopen.com/chapters/68970

Goddek, S., & Korner, O. (2019). A fully integrated simulation model of multi-loop aquaponics: A case study for system sizing in different environments. *Agricultural Systems, 171*, 143–154. https://doi.org/10.1016/j.agsy.2019.01.010

Goddek, S., Joyce, A., Kotzen, B., & Dos-Santos, M. (2019). Aquaponics and global food challenges. In S. Goddek, A. Joyce, B. Kotzen, & G. M. Burnell (Eds.), *Aquaponics food production systems: Combined aquaculture and hydroponic production technologies for the future* (pp. 3–17). Springer. https://link.springer.com/content/pdf/10.1007%2F978-3-030-15943-6.pdf

Gordon, J. M., & Polle, J. E. (2007). Ultra-high bioproductivity from algae. *Applied Microbiology and Biotechnology, 76*(5), 969–975. https://doi.org/10.1007/s00253-007-1102-x

Grand View Research. (2021). *3D bioprinting market size, share & trends analysis report by technology (magnetic levitation, inkjet-based), by application (medical, dental, biosensors, bioinks), by region, and segment forecasts, 2021–2028*. https://www.grandviewresearch.com/industry-analysis/3d-bioprinting-market

Grin, J., Rotmans, J., & Schot, J. (Eds.). (2010). *Transitions to sustainable development: New directions in the study of long-term transformative change*. Routledge.

Herrero, M., Thornton, P. K., Power, B., Bogard, J. R., Remans, R., Fritz, S., Gerber, J. S., Nelson, G., See, L., & Waha, K. (2017). Farming and geography of nutrient production for human use: A transdisciplinary analysis. *Lancet Planetary Health, 1*(1), E33–E42. https://doi.org/10.1016/S2542-5196(17)30007-4

Herrero, M., Thornton, P. K., Mason-D'Croz, D., Palmer, J., Benton, T. G., Bodirsky, B. L., … West, P. C. (2020). Innovation can accelerate the transition towards a sustainable food system. *Nature Food, 1*, 266–272. https://doi.org/10.1038/s43016-020-0074-1

Hoffman, T. (2020). *3D printing: What you need to know?* PC Magazine, Australia. https://au.pcmag.com/3d-printers/41064/3d-printing-what-you-need-to-know

Hydro Naturals. (2021). *Hydroponics*. https://hydronaturals.com/Hydroponics.aspx

Joyce, J. (2013). *All about tillandsia, aka air plants*. The Barn Nursery and Landscape Centre. https://digitblog.barnnurserylandscape.com/blog/bid/238951/All-About-Tillandsia-aka-Air-Plants

Joyce, A., Goddek, S., Kotzen, B., & Wuertz, S. (2019). Aquaponics: Closing the cycle on limited water, land and nutrient resources. In S. Goddek, A. Joyce, B. Kotzen, & G. M. Burnell (Eds.), *Aquaponics food production systems: Combined aquaculture and hydroponic production technologies for the future* (pp. 19–34). Springer. https://link.springer.com/content/pdf/10.1007%2F978-3-030-15943-6.pdf

Kah, M. (2015). Nanopesticides and nanofertilizers: Emerging contaminants or opportunities for risk mitigation? *Frontiers in Chemistry, 3*, 64. https://doi.org/10.3389/fchem.2015.00064

Kasozi, N., Abraham, B., Kaiser, H., & Wilhelmi, B. (2021). The complex microbiome in aquaponics: Significance of the bacterial ecosystem. *Annals of Microbiology, 71*, 1. https://doi.org/10.1186/s13213-020-01613-5

Keating, B. A., Herrero, M., Carberry, P. S., Gardner, J., & Cole, M. B. (2014). Food wedges: Framing the global food demand and supply challenge towards 2050. *Global Food Security, 3*, 125–132. https://doi.org/10.1016/j.gfs.2014.08.004

Kotzen, B., Emerenciano, M. G. C., Moheimani, N., & Burnell, G. M. (2019). Aquaponics: Alternative types and approaches. In S. Goddek, A. Joyce, B. Kotzen, & G. M. Burnell (Eds.), *Aquaponics food production systems: Combined aquaculture and hydroponic production technologies for the future* (pp. 301–330). Springer. https://link.springer.com/content/pdf/10.1007%2F978-3-030-15943-6.pdf

Koyande, A. K., Chew, K. W., Rambabu, K., Tao, Y., Chu, D.-T., & Show, P.-L. (2019). Microalgae: A potential alternative to health supplementation for humans. *Food Science and Human Wellness, 8*(1), 16–24. https://doi.org/10.1016/j.fshw.2019.03.001

Lakhiar, I. A., Gao, J., Syed, T. N., Chandio, F. A., & Buttar, N. A. (2018). Modern plant cultivation technologies in agriculture under controlled environment: A review on aeroponics. *Journal of Plant Interactions, 13*(1), 338–352. https://doi.org/10.1080/17429145.2018.1472308

Lawson, B., & Kaya, M. (2018). *Vertical farming: From Babylon to New York*. Cambridge Consultants. https://www.cambridgeconsultants.com/insights/opinion/vertical-farming-babylon-new-york

Leblanc, R. (2020). *What you should know about vertical farming? Is it the future of agriculture?* The Balance Small Business. https://www.thebalancesmb.com/what-you-should-know-about-vertical-farming-4144786

Lenniy, D. (2021). *Artificial intelligence in agriculture: Rooting out the seed of doubt.* Intellias. https://www.intellias.com/artificial-intelligence-in-agriculture/

Lovell, S. T., Dupraz, C., Gold, M., Jose, S., Revord, R., Stanek, E., & Wolz, K. J. (2017). Temperate agroforestry research: Considering multifunctional woody polycultures and the design of long-term field trials. *Agroforestry Systems, 92*, 1397–1415. https://doi.org/10.1007/s10457-017-0087-4

Marinova, D., & Bogueva, D. (2019). Planetary health and reduction in meat consumption. *Sustainable* Earth, *2*(3). https://doi.org/10.1186/s42055-019-0010-0

Marinova, D., Hong, J., Todorov, V., & Guo, X. (2017). Understanding innovation for sustainability. In J. Hartz-Karp & D. Marinova (Eds.), *Methods for sustainability research* (pp. 217–230). Edward Elgar.

Massy, C. (2017). *Call of the reed warbler: A new agriculture – A new earth.* University of Queensland Press.

McClements, D. J. (2019). *Future foods: How modern science is transforming the way we eat.* Springer.

McClements, D. J. (2020). Future foods: Is it possible to design a healthier and more sustainable food supply? *Nutrition Bulletin, 45*, 341–354. https://doi.org/10.1111/nbu.12457

McClements, D. J., Barrangou, R., Hill, C., Kokini, J. L., Lila, M. A., Meyer, A. S., & Yu, L. (2021). Building a resilient, sustainable, and healthier food supply through innovation and technology. *Annual Review of Food Science and Technology, 12*, 1–28. https://doi.org/10.1146/annurev-food-092220-030824

Meaning Directory. (2016). *Meaning of "ponic"*. https://meaningdir.com/word/ponic.htm

Meghani, N., Dave, S., & Kumar, A. (2020). Introduction to nanofood. In U. Hebbar, S. Ranjan, N. Dasgupta, & R. Kumar Mishra (Eds.), *Nano-food engineering* (pp. 1–23). Springer. https://doi.org/10.1007/978-3-030-44552-2_1

Meisenzahl, M. (2020). *A startup is 3D printing plant-based steaks to recreate the taste and texture of the real thing – See how they do it.* Business Insider Australia. https://www.businessinsider.com.au/redefine-meat-3d-printed-plant-based-faux-steaks-in-photos-2020-9?r=US&IR=T

Meyer, M. (2017). *The evolution of "ponics" in organic.* Organic Matters Blog. https://organic-mattersblog.com/2017/07/06/the-evolution-of-ponics-in-organic/

Mishra, S., Keswani, C., Abhilash, P. C., Fraceto, L. F., & Singh, H. B. (2017). Integrated approach of agri-nanotechnology: Challenges and future trends. *Frontiers in Plant Science, 8*, 471. https://doi.org/10.3389/fpls.2017.00471

Muller, A., Ferré, M., Engel, S., Gattinger, A., Holzkämper, A., Huber, R., Müller, M., & Six, J. (2017). Can soil-less crop production be a sustainable option for soil conservation and future agriculture? *Land Use Policy, 69*, 102–105. https://doi.org/10.1016/j.landusepol.2017.09.014

NASA Spinoff. (2006). *Progressive plant growing has business blooming.* https://spinoff.nasa.gov/Spinoff2006/er_2.html

National Aeronautics and Space Administration (NASA). (2007). *Progressive plant growing is a blooming business.* https://www.nasa.gov/vision/earth/technologies/aeroponic_plants.html

Panwar, A. (2020). *The origins of hydroponics.* I Grow. https://www.igrow.news/igrownews/the

Park, W. (2020). *The farms growing beneath our cities.* BBC, Follow the Food. https://www.bbc.com/future/bespoke/follow-the-food/the-massive-farms-emerging-beneath-our-cities.html

Parkinson, E. (2016). *Agriculture goes vertical as buildings become the new farms.* Financial Review. https://www.afr.com/companies/agriculture-goes-vertical-as-buildings-become-the-new-farms-20160216-gmv7z8

Poore, J., & Nemecek, T. (2018). Reducing food's environmental impacts through producers and consumers. *Science, 360*, 987–992. https://doi.org/10.1126/science.aaq0216

Pure Greens Arizona. (2020). *Richard J. Stoner – The father of modern aeroponics*. https://medium.com/@PureGreensAZLLC/richard-j-stoner-the-father-of-modern-aeroponics-58e7f4ac6193

Qazi, G., & Dar, F. A. (2020). Nano-agrochemicals: Economic potential and future trends. In U. Hebbar, S. Ranjan, N. Dasgupta, & R. Kumar Mishra (Eds.), *Nano-food engineering* (pp. 185–193). Springer. https://doi.org/10.1007/978-3-030-39978-8_11

Ramachandraiah, K. (2021). Potential development of sustainable 3d-printed meat analogues: A review. *Sustainability, 13*, 938. https://doi.org/10.3390/su13020938

Rodale Institute. (2014). *Regenerative organic agriculture and climate change: A down-to-earth solution to global warming*. https://rodaleinstitute.org/wp-content/uploads/rodale-white-paper.pdf

Sankhalkar, S., Komarpant, R., Dessai, T. R., Simoes, J., & Sharma, S. (2019). Effects of soil and soil-less culture on morphology, physiology and biochemical studies of vegetable plants. *Current Agriculture Research Journal, 7*(2), 181–188. https://doi.org/10.12944/CARJ.7.2.06

Sathasivam, R., Radhakrishnan, R., Hashem, A., & Abd Allah, E. F. (2019). Microalgae metabolites: A rich source for food and medicine. *Saudi Journal of Biological Sciences, 26*(4), 709–722. https://doi.org/10.1016/j.sjbs.2017.11.003

Sekhon, B. S. (2014). Nanotechnology in agri-food production: An overview. *Nanotechnology, Science and Applications, 7*, 31–53. https://doi.org/10.2147/NSA.S39406

Sheehan, M. (2019). *How drones are being used in Australia to make farming more efficient*. Foresttech. https://foresttech.events/how-drones-are-being-used-in-australia-to-make-farming-more-efficient/

Silva, G., & Di Serio, L. C. (2016). The sixth wave of innovation: Are we ready? *RAI Revista de Administração e Inovação, 13*(2), 128–134. https://doi.org/10.1016/j.rai.2016.03.005

Smith, A., Stirling, A., & Berkhout, F. (2005). The governance of sustainable socio-technical transitions. *Research Policy, 34*(10), 1491–1510. https://doi.org/10.1016/j.respol.2005.07.005

Steffen, W., Richardson, K., Rockström, J., Cornell, S. E., Fetzer, I., Bennett, E. M., Biggs, R., Carpenter, S. R., de Vries, W., de Wit, C. A., Folke, C., Gerten, D., Heinke, J., Mace, G. M., Persson, L. M., Ramanathan, V., Reyers, B., & Sörlin, S. (2015). Planetary boundaries: Guiding human development on a changing planet. *Science, 347*(6223), 1259855. https://doi.org/10.1126/science.1259855

Stoner, R. J. (1983). Aeroponics versus bed and hydroponic propagation [The process of propagating and growing plants in air]. *Florists' Review, 173*(4477), 49–51.

Thomaier, S., Specht, K., Henckel, D., Dierich, A., Siebert, R., Freisinger, U. B., & Sawicka, M. (2015). Farming in and on urban buildings: Present practice and specific novelties of Zero-Acreage Farming (ZFarming). *Renewable Agriculture and Food Systems, 30*(1), 43–54. https://doi.org/10.1017/S1742170514000143

Time. (1937). *Science: Hydroponics*. http://content.time.com/time/subscriber/article/0,33009,757343,00.html

Tirado, R., Thompson, K. F., Miller, K. A., & Johnston, P. (2018). *Less is more: Reducing meat and dairy for a healthier life and planet*. Greenpeace Research Laboratories Technical Report (Review) 03-2018. https://www.greenpeace.org/international/publication/15093/less-is-more/

Torabi, M., Mokhtarzadeh, A., Mahlooji, M., & Iran, P. (2012). The role of hydroponics technique as a standard methodology in various aspects of plant biology researches. In T. Asao (Ed.), *Hydroponics: A standard methodology for plant biological researches* (pp. 113–134). IntechOpen. https://www.intechopen.com/books/1781

United Nations (UN). (2019). *World population prospects 2019: Highlights* (ST/ESA/SER.A/423). Department of Economic and Social Affairs, Population Division. https://population.un.org/wpp/Publications/Files/WPP2019_Highlights.pdf

United Nations (UN). (2021). *The Sustainable Development Goals report*. https://unstats.un.org/sdgs/report/2021/The-Sustainable-Development-Goals-Report-2021.pdf

United Nations Conference on Trade and Development (UNCTAD). (2021). *Technology and Innovation Report 2021: Catching technological waves; Innovation with equity.* https://unctad. org/system/files/official-document/tir2020_en.pdf

Van Gorcum, B., Goddek, S., & Keesman, K. J. (2019). Gaining market insights for aquaponi- cally produced vegetables in Kenya. *Aquaculture International, 27,* 1231–1237. https://doi. org/10.1007/s10499-019-00379-1

Van Krimpen, M. M., Bikker, P., van der Meer, I. M., van der Peet-Schwering, C. M. C., & Vereijken, J. M. (2013). *Cultivation, processing and nutritional aspects for pigs and poul- try of European protein sources as alternatives for imported soybean products.* Report 662. Wageningen University and Research. https://edepot.wur.nl/250643

Van Loey, A., Foubert, I., Gheysen, L., & Bernaerts, T. (2020). *Microalgae: Food of the future?* New Food. https://www.newfoodmagazine.com/article/118441/microalgae-food-of-the-future/

Wanger, T. C., DeClerck, F., Garibaldi, L. A., Ghazoul, J., Kleijn, D., Klein, A.-M., … Weisser, W. (2020). Integrating agroecological production in a robust post-2020 global biodiver- sity framework. *Nature Ecology & Evolution, 4,* 1150–1152. https://doi.org/10.1038/ s41559-020-1262-y

Weiseman, W., Halsey, D., & Ruddock, B. (2014). *Integrated forest gardening: The complete guide to polycultures and plant guilds in permaculture systems.* Chelsea Green Publishing.

Willett, W., Rockström, J., Loken, B., Springmann, M., Lang, T., Vermeulen, S., … Murray, C. J. L. (2019). Food in the Anthropocene: The EAT–Lancet Commission on healthy diets from sustainable food systems. *The Lancet, 393*(10170), 447–492. https://eatforum.org/con- tent/uploads/2019/01/EAT-Lancet_Commission_Summary_Report.pdf

World Agroforestry. (2021). *What is agroforestry?* https://www.worldagroforestry.org/about/ agroforestry

World Bank. (2018a). *Agricultural land (% of land area).* https://data.worldbank.org/indicator/ ag.lnd.agri.zs

World Bank. (2018b). *Arable land (% of land area).* https://data.worldbank.org/indicator/ AG.LND.ARBL.ZS

Xiong, H., Dalhaus, T., Wang, P., & Huang, J. (2020). Blockchain technology for agriculture: Applications and rationale. *Frontiers in Blockchain, 3,* 7. https://www.frontiersin.org/ articles/10.3389/fbloc.2020.00007/full

Xu, J., Guo, S., Xie, D., & Yaxuan, Y. (2020). Blockchain: A new safeguard for agri-foods. *Artificial Intelligence in Agriculture, 4,* 153–161. https://doi.org/10.1016/j.aiia.2020.08.002

Youmatter. (2020). *What is agroecology? What is the history behind it? What are some examples of agroecology? What are the principles of agroecology? Let's find out.* https://youmatter.world/ en/definition/definitions-agro-ecology/

Zhang, Z., & Wang, X. J. (2017). Current progresses of 3D bioprinting based tissue engineering. *Quantitative Biology, 5,* 136–142. https://doi.org/10.1007/s40484-017-0103-8

Chapter 7
Alternative Proteins

Abstract The term "alternative proteins" describes alternatives and substitutes to animal-based foods. This area is gaining popularity fast as humanity tackles the challenges of climate change, biodiversity loss, land- and sea-use changes and food security. Some alternative protein sources have existed for millennia, including tofu, tempeh and seitan, but others, such as cultured meat and 3D food, are new. The chapter follows a historical line in these developments analysing the pros and cons of alternative proteins and offering reflections about their future, potential consumers and challenges.

The human body cannot store proteins and requires daily intake of essential amino acids to maintain good health. Animal-based proteins are a good and easy source of these essential amino acids but so are some plants that can be used for food, such as soya, buckwheat, chia, quinoa, industrial hemp (i-hemp) seeds and chickpeas. They are described as complete protein sources. Furthermore, a simple combination between beans and rice can also provide the complete set of essential amino acids required by the human body (Marshall & Marinova, 2019). The range of sources of proteins alternative to animal-based foods expands to include insects, fungi and the newest innovations in lab-grown (cultured or clean) meat. A further group are the analogues or new plant-sourced foods which closely resemble standard livestock-based options, such as mince, sausages, milk, eggs or mayonnaise. This chapter analyses the various alternative proteins discussing the progress made in their acceptance as mainstream foods and as business opportunities.

In a visit to China in 2018, we were astonished and amazed by the diversity of insects offered in multiple forms and preparation methods in the main street food market in Shanghai (see Fig. 7.1). Scorpions live and cooked, pierced on a stick, silkworm larvae, mealworms, cicadas, crickets, various types of known and unknown bugs, all were available there to please someone's palate. We watched with curiosity when locals passed them unabashedly. Foreign tourists like us were the only customers who dared to touch or taste the bugs as an exotic experience, accompanied with an authentic photo. With prodigious curiosity, we asked one of

D. Marinova, D. Bogueva, *Food in a Planetary Emergency*,
https://doi.org/10.1007/978-981-16-7707-6_7

Fig. 7.1 Shanghai Food Market. (Photo: Authors)

the stall holders why there was limited interest by local people for what we believed was a common food option. The tradition to eat insects in China dates back to more than 3000 years ago. This was confirmed by the stall holder who also told us that this practice is customary predominantly among the Chinese minority groups. One-hundred and seventy-eight insect species from 96 genera, 53 families and 11 orders are commonly eaten in China (Chen et al., 2009). However, these are no longer considered a substantial part of the Chinese diet, which has now turned its utter attention to Western foods rich in meat. The trends reflecting this shift are evident in the world meat consumption statistics. In the last 20 years up to 2018, meat consumption per person in China has exponentially grown by 54%, compared to lower increases in Australia (13%) and USA (8%) – the world's top meat-consuming countries (Whitnall & Pitts, 2019).

Why do we start this chapter with insects as a source of protein? The answer is simple – they are the oldest, like humankind itself, alternative protein sources on Earth. Insects have been an integral part of our ancestral hunter-gatherer lifestyle and foraging past, and were eaten from tens of thousands of years ago. This is still practised by more than 2 billion people around the world, including the Australian Aboriginals, as a sustainable source of food. Despite the usually provoked negative thoughts and the yuck factor attached to them for the western consumer, something is fascinating about insects. Today over 1500 species of insects are regularly consumed by humans from over 300 ethnic groups in 113 countries around the world (MacEvilly, 2000).

Amongst Westerners nowadays, the thought about eating insects is engulfed in many connotations and emotions of revolting sensation, disgust, neophobia, nausea and other negative feelings. Despite being at ease with eating the sea cousins of insects – prawns, shrimps or lobsters, the psychological rejection of their terrestrial counterparts revolves around reasons related not only to food. There is unacceptance of ethnic and traditional habits and customs that are viewed as foreign and therefore unknown, unfamiliar and strange. Eating insects is associated with a creepy-crawly feeling (Sogari et al., 2019) but also with the processes related to the decomposition of a decaying dead body where maggots help recycle carrion back into nature. Such perceptions result in food disgust closely connected with the idea of something pathogenic that frequently triggers behavioural avoidance (Hartmann & Siegrist, 2020). These perceptions lie on the border between our sense of self and otherness, our ideas about good and bad, health and illness, pleasant and nasty, taste and distastefulness (Miller, 2014). For many in the affluent world, insects are part of a genuine feeling of aversion.

As a mechanism for food rejection experienced consciously or unconsciously by humans, the feelings of disgust and aversion are not new. The word "disgust" which now expresses repugnance at rotten food or filth, was unknown during the Shakespeare's era and the feeling was explained as making someone's gorge raise (Merriam-Webster, n.d.). In 1590–1600, the world entered into English from the Old French "desgouster" (nowadays dégoûter), originating from Latin – des- or dis- for apart and goust for taste (Dictionary.com, n.d.). The meaning of disgust or dislike is referent to the sense of bad taste in the mouth (Lane, 2011). In the eighteenth century with the advent of the Industrial Revolution, it became common amongst journalists, politicians and commentators to employ "disgust" and its derivatives (e.g. disgusted, disgustful) to describe not reactions to food, but opposition to things they did not like. In more recent times, disgust started to be used to show fear of toxicity and disease (Chapman & Anderson, 2012) and to express negative emotional reactions to a particular product's attributes representing barriers to consumption (Rozin & Fallon, 1987). Most importantly, people's food disgust shapes their eating preferences and behaviour, including food waste (Egolf et al., 2018).

In Shanghai, we encountered a sense of disgust, especially when some brave tourists put and swallowed different insects in their mouths. Our previous research with Australia's Millennials and Generation Z (Sogari et al., 2019) showed similar negative feelings in relation to insects being used as food. These young consumers otherwise open to new experiences (Deloitte, 2021), were not keen to include insects in their meals. They demonstrated low willingness to accept insects as a substitute for meat because of strong psychological barriers, such as neophobia and disgust, combined with concerns about threats to masculinity and limited understanding of any environmental and nutritional benefits (Sogari et al., 2019). Traditional insect consumption is unlikely to become a widespread practice in mainstream Western culture, at least for now.

At the 2019 Cutting Edge Symposium "Developing Australia's edible insect research and industry to improve environmental, health and cultural outcomes" organised by the Commonwealth Scientific and Industrial Research Organisation (CSIRO) and representatives from the emerging edible insects industry, there was a question from the audience after our presentation: "When will Australian consumers accept eating insects?" The answer was straight to the core: "It is unlikely to happen in the next 10 years, but fortunately for the planet people will pursue other dietary changes in the search for alternatives to the livestock proteins and this will help their health and the health of the planet". It is important to have a sense of reality and this chapter provides an overview of such alternative proteins.

How Novel Are the Alternative Proteins

Although the term "alternative proteins" is relatively new, it is fast gaining popularity as humanity tackles the challenges of climate change, biodiversity loss, land- and sea-use changes and food security. The implied alternativeness or substitution is for proteins which do not originate from animals, particularly livestock and fish. According to CSIRO (Wynn & Sebastian, 2019), alternative proteins describe foods consumed as substitutes to meat and seafood in a person's diet. Another implied characteristic of these alternative proteins is that they need to contain the full range or a substantial part of the essential amino acids (histidine, isoleucine, leucine, lysine, methionine, phenylalanine, threonine, tryptophan and valine) which the human body cannot produce itself and have to come from food. By comparison, the non-essential amino acids (alanine, arginine, asparagine, aspartic acid, cysteine, glutamic acid, glutamine, glycine, proline, serine and tyrosine) can be produced by the human body by combining components from the food we eat (except when we are ill or stressed).

Such alternative protein sources have existed for millennia, with products made traditionally from soybeans, such as tofu and tempeh, and from wheat protein, such as seitan. They have been and continue to be used as affordable, functional and nutritious protein sources (Lawrence & King, 2019). Previously forgotten or dismissed, they have become over the past decade an emerging field of interest, discussions, study, research, development and innovation, and a major conversation topic within the agrifood industry, attracting multibillion-dollar investments. The particular focus on these alternative proteins is because of issues of food security as today's food systems need to feed a global population of 7.9 billion and in the future, the 10 billion expected by 2050. The Earth is rapidly approaching its productive and resource limits. Furthermore, consumers' interest is increasingly shaped by three major drivers – the desire for healthy foods; sustainability, including sustainable sources of proteins; and ethical considerations.

At present, the most widely available plant-based alternative proteins for human consumption are produced from textured soya and pea protein, mycoprotein from mushrooms and wheat gluten. Other sources currently being explored are the application of lupines, beans, rice, algae and jackfruit, although many are at an early development stage. Cultured meat is seemingly drawing a lot of attention and investment and so are the plant-based dairy products. Replacement of seafood is also part of these new developments. Let's look at them in turn starting as we already did, with insects. We also try to follow a historical line in these developments, although this is not always simple because of the various interactions, cultural and trade exchanges across the globe. The pros and cons of the alternative proteins are then discussed followed by some reflections about their future, potential consumers and challenges.

Edible Insects

There are developments in the field of edible insects which may change our perspective about their unpopularity. Insects are currently underestimated and undervalued in their potential to address global challenges, such as dealing with climate volatility, feeding a growing population and easing the pressures on natural resources. Until recently insects were primarily sourced from the wild but nowadays farming them as animal feed, for waste management and for human consumption is on the rise. Despite some current challenges in scaling-up production, farmed insects are starting to play a role in our global food systems. Insectile protein is regarded as the most efficient in the conversion of feed into edible weight (2.3 kg of feed for 1 kg of edible weight) and can be raised on low-value agricultural by-products (Bashi et al., 2019). Crickets, being the most common edible insect and a good source of protein, are already milled for flour by some producers, but on a smaller scale due to the low cost ineffectiveness of cricket-based protein production. Similarly, food producers are exploring the use of grasshoppers as an edible insect source, but development in this area is still at an early stage (Watson, 2019). The unfamiliar taste, texture and aroma of insects still require extensive research and development together with addressing safety perceptions related to potential allergens (Bashi et al., 2019).

Despite the range of negative feelings and views, projections show the global edible insects market to reach near $8 billion by 2027 (Meticulous Research, 2019). This growth is primarily motivated by population increase and declining food resources, growing demand for protein-rich foods, high prices of animal proteins, better environmental sustainability, the high nutritional value of insects and low risk of transmitting zoonotic diseases. Nevertheless, the global non-harmonised regulatory framework, information and awareness paucity, psychological and ethical barriers and potential allergies are key issues that limit the development of the edible insects market. While traditionally people in Africa will continue to eat caterpillars, crickets, palm weevils and termites without considering them being pests or a nuisance (Niassy & Ekesi, 2017), and Australian Aborigines indulge on whitchetty grub – the wood-eating larvae of the cossid moth, the growing global population will have to reposition its perspective on what is safe and tasty to eat.

Tofu

Tofu was created from soybeans during the Han dynasty dating back to 206 BCE–9 CE (Tan, 2008; Du Bois, 2018; Lawrence & King, 2019). Prince Liu An, the ruler of the Huainan Kingdom, is believed to have invented the bean curd method of making tofu as well as soy milk (see Fig. 7.2). A Tao Gu (903–970) written source describes tofu as the "vice mayor's mutton" (Shurtleff & Aoyagi, 2013) alluding to the special value of the soy-based products as an affordable alternative to meat. Tofu making involves peculiar processing starting with soaking and then milling the soybeans, followed by filtering and cooking of the produced soymilk, stirring the coagulant and finishing the process with pressing the curd to drain any excess water and ferment (see Fig. 7.2). The entire process is depicted in illustrations on the Han dynasty tomb (Tan, 2008) affirming the significance of tofu as a distinctive and important food product.

This valuable food product was introduced progressively to Japan, Korea and Vietnam. In 1603, tofu was first mentioned in a European language in the Portuguese

Fig. 7.2 Prince Liu An and Tofu making. (Source: Authors)

document "Vocabulario da lingua de Iapam" (Vocabulary of the language of Japan) and in 1880, was first made in a western country, France (Shurtleff & Aoyagi, 2013). Since 1905, tofu has been available on the American market while its commercial production in USA commenced in 1929. By 1982, there were 242 tofu manufacturers in the western world and in 1999, the US Food and Drug Administration (FDA) authorised the claim that a consumption of 6.25 g per serving lowers cholesterol and reduces the risk of heart disease (Shurtleff & Aoyagi, 2013). Nowadays tofu is a widely accepted and popular product worldwide and is available in diverse types, such as soft, silken or firm. It is a well-established alternative protein containing all nine essential amino acids as well as being a good source of iron, calcium, manganese, phosphorus, magnesium, copper, zinc and vitamin B1. Moreover, being a legume, the soybeans help maintain soil fertility through nitrogen fixation.

The popularity of soybeans is not only for human consumption. They are also used as animal feed fuelling an ever-increasing demand for the "miracle crop" (Guo et al., 2019) with a technological response through genetically modified organism (GMO) varieties. In 2017, 68% of the 1.1 billion tonnes of soybeans for domestic consumption in China were used as animal feed while the Chinese diet was becoming increasingly westernised (Guo et al., 2019). Many good recipes and healing properties of soyfoods preserved throughout the millennia (Shurtleff et al., 2014) are progressively being forgotten. The inefficient use of resources to feed livestock instead of directly people, has resulted in 82% of the world's soybean farms relying on GMO crops (Wong & Chan, 2016) putting soy-based plant products under scrutiny.

Tempeh

Tempeh, which originated from central or east Java between a few centuries to a thousand years ago, continues to be one of the most popular fermented foods in Indonesia (Shurtleff & Aoyagi, 2007). Made from fermented soybeans, filamentous fungi (such as *Rhizopus* and *Penicillium*) play an important role in the fermentation process. Despite its old origin, the first written mention of the earthy, nutty smelling, firm and moist in texture tempeh is in the early 1800s (Shurtleff & Aoyagi, 2007). Tempeh began to be commercially produced in Europe between 1946 and 1959, and by 1984 there were 18 European tempeh companies, some of them started in the Netherlands by immigrants from Indonesia. The first English-language reference to tempeh was by Stahel who in 1946 was the director of the Agricultural Experiment Station in Suriname – a tropical country on the north-eastern Atlantic coast of South America administered at the time by USA under an agreement with the Dutch government. In the early 1960s, extensive research on tempeh took place at Cornell University and since 1961 the soybean-based food has been commercially produced in USA, initially by Indonesian immigrants (Shurtleff & Aoyagi, 2007).

In addition to soybeans, tempeh can also be made from other bean varieties and from cereal grains, such as wheat, sorghum, barley and rye, as well as a mixture between them (Hachmeister & Fung, 1993). Following the fermentation, which helps develop the desirable texture, aroma and flavour, the beans/grains are joined together in a compact cake by dense cottony mycelium. In Australia, the interest in tempeh began in 1977 after research published by McComb on the use of sweet narrow-leafed lupins (Shurtleff & Aoyagi, 2007). Due to its compactness, soy-based tempeh is higher in proteins compared to tofu and it also contains iron, calcium, magnesium, phosphorus, manganese, riboflavin and niacin. Tempeh consumption is associated with increased production of antibodies, reduced risk of diabetes and lower cholesterol levels (Mordor Intelligence, 2021) making it an attractive alternative protein source.

The current market for tempeh is shaped by many converging trends, including the developing demand for ethically sourced food, healthier options, probiotics and savoury taste, increasing gluten intolerance as well as lower cost compared to other meat substitutes (Mordor Intelligence, 2021). Although North America currently dominates the market, demand in the Asia-Pacific region is fast-growing driven by countries, such as China, South Korea, Indonesia, Japan and Australia (Hexa Research, 2019; Mordor Intelligence, 2021). The global supply of tempeh is fragmented and diversified with many companies operating at a local level (Hexa Research, 2019). Demand for this alternative protein is growing in response to the planetary emergency.

Seitan

The wheat protein seitan, made entirely out of hydrated gluten, originated in China in the sixth century (McGuiness, 2019) and was widely consumed as a substitute for meat, especially among followers of Buddhism (Anderson, 2014). The word seitan is of Japanese origin describing "made of protein" and was coined in 1961 by George Ohsawa, a Japanese advocate of the macrobiotic diet.

Seitan has been used as a meat substitute for centuries in China and Japan. It is produced by removing the starch from the wheat rinsing it with water. Seitan is believed to have chicken-like texture, fibrous and elastic, also described as toothsome, and is used as an ingredient in many mock meat products. It could mimic meat's flavour, heartiness and has the rare qualities to absorb sauces (Bramen, 2010; Mistry et al., 2020).

More recent varieties of seitan can use wheat gluten and chickpeas (McGuiness, 2019). As seitan is made of wheat, it does not contain the essential amino acid lysine and in this sense, it is not considered a complete protein, however this problem can be avoided by mixing it with chickpeas or beans in the production process or in meals during consumption. A variety of spices are also added, including to imitate the taste of animal meat-based products, such as sausages.

Meatless Meat

In medieval Europe, animal meat was expensive and prestigious (Bogueva et al., 2018). It was common to create dishes which pretended to imitate meat foods (Adamson, 2004). People, and especially the noble, used to indulge in chopped grapes and almonds and diced bread as a substitute for minced animal-based meat during Lent when the consumption of warm-blooded animals, eggs and dairy products was prohibited. Diced bread was often made into imitation cracklings and greaves. In late medieval Europe, imitation also displayed playfulness where "the idea of playing with food was elevated to an art form" (Adamson, 2004, p. 71).

From abstaining because of religious rules and economic unaffordability, the "meatless meat" practice shifted along nutritional considerations. Around 1896, the supporter of the holistic, biologic living approach and nutrition pioneer Dr. John Harvey Kellogg developed America's first peanut, grain and soy-based meat analogue. The move toward creating a "vegetable meat" was suggested by the then Assistant Secretary of Agriculture at the United States Department of Agriculture (USDA) and a renowned agricultural chemist Dr. Charles W. Dabney, who also recommended the use of navy beans for this purpose (Shurtleff & Aoyagi, 2004). Kellogg believed the new plant-sourced meats, such as Nuttose or Protose, to be healthier substitutes for animal-based meat to feed patients providing them with all the necessary proteins. The Sanitarium hospital was established first in Michigan, USA and then in Sydney, Australia (now Sydney Adventist Hospital), based on the teachings of the Seventh-day Adventist Church where "doctors advocated healthier eating, drinking fresh water and exercising outdoors to help people 'learn to stay well' (the meaning of the word Sanitarium)" (Sanitarium, 2021, para. 2). Depending on the meal purpose, the new meatless meat could be seasoned to taste like beef, veal, chicken or salmon, and served sliced or diced (Schwartz, 1970). The not-for-profit Sanitarium company was established by Dr. John Harvey Kellogg in Melbourne in 1898 while his brother Will Keith Kellogg founded the Kellogg's cereal company in Battle Creek, Michigan, USA in 1906 (Praestiin, 2020). Nowadays, these two companies continue to support plant-based products, including plant-based meat alternatives.

Forty years later, in 1937, Dr. Kellogg also created the first acidophilus soymilk and a soy acidophilus cheese (Shurtleff & Aoyagi, 2004). The soy-based cottage cheese claimed to contain 250,000 million L. acidophilus organisms per gram. Around the same time, in 1931 Winston Churchill (1931, para. 11) wrote: "We shall escape the absurdity of growing a whole chicken to eat the breast or wing, by growing these parts separately under a suitable medium" predicting the emergence almost 90 years later of cultured meat which we discuss later in this chapter.

During late nineteenth century and the beginning of twentieth century, the interest in meat analogues increased, particularly by vegetarians who were looking for alternatives to animal proteins due to ethical and cultural reasons (Spencer, 1993) connected to non-violence towards animals and the Indian philosophies of Hinduism and Jainism (see Fig. 7.3). Cookbooks with meat alternative recipes, such as *Mrs*

Fig. 7.3 Monument to the founder of the Society for the Prevention of Cruelty to Animals, Kolkata, India. (Source: Authors)

Rorer's Vegetable Cookery and Meat Substitutes published by the dietician Sarah Tyson Rorer in 1909, advocated their health benefits. One of the recipes was mock veal roast made from lentils, breadcrumbs and peanuts (Shprintzen, 2013). During and after the two World Wars, regular meat-eaters were confronted with food shortages, significantly reduced access to meat and were also looking for alternative proteins (Perren, 2006). Any available beef or pork was sent to the soldiers at the front or given to the working men as a good source of energy (Collingham, 2011). Many recipes in cookbooks from that time offered a variety of replacements for meats and dairy products, including mock goose, duck, crab and cream made from readily available plant ingredients, such as potatoes, carrots, root vegetables (celeriac, squashes and gourds) and onions (BBC, n.d.). In 1945, the American pioneer of natural and health foods Mildred Lager known for popularising soy-based recipes (Nguyen, 2012), commented that soybeans "are the best meat substitute from the vegetable kingdom, they will always be used to a great extent by the vegetarian in place of meat" (Lager, 1945, p. 106).

By 1954 meat became more available in Europe (BBC, n.d.), but since the late 1960s concerns about feeding the growing world population on animal protein alone have persisted. The issues about food production and consumption have been further augmented by the current planetary emergency stressing the importance of alternative proteins. Initially, the meat alternatives were limited to plant-sourced options, such as soy- and nut-based meats, with no reminiscence of conventional animal protein in relation to texture, taste, appearance and some, especially soy-based, products having unpleasant odour (Mistry et al., 2020). Today the food industry has improved these options to new consumer satisfactory levels. The current technologies for producing meat analogues, including thermoplastic extrusion and fibre spinning processes, not only provide the food industry with solutions-driven instruments, but also with cost-effective methods of accommodating large-scale production and tools for forming the consumer-desired fibres (Alam et al., 2016). With these technological advances, the food industry is working on creating analogues with superior functionality and sensory attributes to the ones of the animal-based products (Beniwal et al., 2021). Let's look at some of the more recent achievements.

Textured Vegetable Protein (TVP)

Textured (or texturised) Vegetable Protein (TVP) is produced from soy flour or concentrate with 50–70% soy protein and is a by-product of soybean oil extraction. It is extruded under high-moisture and high-temperature conditions into various shapes – granules, nuggets, chunks (Groves, 2018) to form fibrous, insoluble, spongy substance which has a long shelf life and can soak up liquids when cooked. This vegan meat substitute was developed by the food conglomerate Archer Daniels Midland in the 1960s (Mistry et al., 2020). It was a creative way to use waste products known as oil-cakes, a left-over from the extraction process of vegetable oils. The oil-cakes can be given to livestock animals, but through the TVP extrusion and processing they can also provide a valuable source of protein for plant-based alternatives for human consumption (Mistry et al., 2020). In 1996, the TVP patent which was held by Krafts Foods Global, expired and now the technology is available in the public domain.

In addition to soy, the application of the high-moisture extrusion process can transform other legume and grain protein from peas, lupines, cottonseed, wheat and oat into a fibrous structure similar to animal meat. These products are palatable and acceptable to consumers (Liu & Hsieh, 2008). Adding seasoning, aromas and colourings makes TVP suitable for creating meat analogues or extenders which make meat products healthier (FAO, 2018).

Quorn

Also in the 1960s, Quorn was developed as an alternative meat product from a fungus growing in Marlow, Buckinghamshire area in UK. In a search for a sustainable source of protein, Lord Rank discovered the naturally occurring fungi family *Fusarium venenatum* which converts carbohydrate into protein-rich mycoprotein through a fermentation process (Finnigan et al., 2019). His persistence in screening over 3000 soil samples from around the world led to this discovery. Twenty years later in the early 1980s, Quorn was created packed with fibre, low in saturated fat, containing no cholesterol, sodium or sugar (Finnigan et al., 2019). According to research evidence, its richness in proteins and fibres is beneficial for retaining a healthy gut microbiome (Mistry et al., 2020), maintaining healthy blood cholesterol levels, promoting muscle synthesis as well as controlling glucose and insulin levels, and it also increases satiety (Finnigan et al., 2019). The standard Quorn options use egg albumen to bind together the dried fungus culture while potato protein is employed as a binder in the vegan versions.

After the patent for the mycoprotein-based Quorn products expired in 2010, other companies can now use the technology (Lindsay, 2018). These products have already gained popularity in UK, the rest of Europe and USA (Apostolidis & McLeay, 2016) due to their quality and versatile advertising targeting both vegetarians and meat-eaters (Mintel, 2013). In UK, consumers associate Quorn with health and environmental benefits driven by values of food security, benevolence and universalism (Apostolidis & McLeay, 2016).

There are some concerns raised in relation to Quorn products. They can cause severe, adverse allergic reaction to mycoproteins in some people (Mistry et al., 2020). Also, in Canada Quorn is banned due to lack of compliance with the Canadian Food Inspection Agency's nutrient content claims requirements; however, it has been approved by Health Canada (Quorn, 2021). Currently, the original Quorn products are sold in 16 countries around the world (Quorn, 2021).

Cultured Meat

More than 90 years ago in 1930, the British Conservative politician Frederick Edwin Smith and later Winston Churchill prophesied that people will no longer need to grow an entire animal to eat part of it, be it steak from a bullock or breast from a chicken (Bhat et al., 2015). This sounded like a fantasy until 5 August 2013 when Professor Marcus Johannes (Mark) Post from Maastricht University made it a reality by growing meat in a lab test tube (Post, 2012). In September 2020 and in a world first, cultured meat was approved and officially authorised as food for the mainstream consumer by the Singapore Food Agency to pave the way for other sustainable food products and better food security (Tan, 2020).

Cultured meat is also described as in-vitro, lab-grown, cell-based, cell-cultured, cultivated, fake or clean meat. It is grown in bioreactors and with investment and production expanding, the term factory-grown meat (Post, 2012) may also gain popularity in the future. When we were researching the attitudes of Generation Z towards cultured meat (Bogueva & Marinova, 2020), however, we realised that young people in Australia have not yet accepted the idea of growing and eating cultured meat (some described it as a Frankenstein idea). Cultured meat was seen more as narrative than tangible, eat-able foodstuffs (Mouat & Prince, 2018). After Singapore's progressive move to approve the sale of cultured meat, it is now real to be touched, tasted, assessed and enjoyed. Perhaps for many other consumers around the world this is not yet a close reality but the intention behind cultured meat is to offer alternative proteins that are more sustainable, safer and pave the way to food security.

There has been a remarkable decrease in the price of cultured meat from the first available patty at a cost of $300,000 in 2013 (Chriki & Hocquette, 2020) to the projected fall to $10–$15 by 2021 (Axworthy, 2019; Tan, 2020). It is also expected to reach parity with regular conventionally farmed meat and become even cheaper. As of 2021, cultured meat is between 100 and 10,000 times more expensive than livestock meat depending on the exact requirements, cell types and use of growth factors and recombinant proteins (CE Delft, 2021b). However, parity can be achieved by 2030 when production costs can reach $5.66/kg (CE Delft, 2021b). Any automation or artificial intelligence (AI) can decrease production costs even further. The factors that drive cost down also decrease the environmental footprint of cultured meat related to energy use, including source and efficiency, as well as efficiency in the production and supply chain (CE Delft, 2021a, b). In 2021 alone, cultured meat attracted $550 million of investment and it is projected that by 2030, the industry will be worth $25 billion globally (Brennan et al., 2021).

The fact that animals do not need to be raised and slaughtered to produce meat for human consumption is appealing to animal rights and welfare groups and sections within society. This is likely to generate further public support, particularly with increased environmental benefits related to reduced requirements for agricultural land and animal feed. A drastic reduction in the current land use by the livestock industry will open up perspectives for reforestation and restoring wildlife in many parts of the globe. The potential for water saving is also immense. Food safety and nutritional benefits can further be achieved by controlling the lab environment, how cultured meat is produced and eliminating the need to use antibiotics. Improvements in energy use, particularly a shift to renewable sources, will give cultured meat bigger environmental advantages. Compared to beef, pork and chicken, the production of cultured meat will reduce the global warming impacts by 85–92%, 52% and 17% and land use by 95%, 72% and 63%, respectively (Watson, 2021).

All these developments were made possible because of Post's revolutionary innovation and creation of his start-up Mosa Meat which was followed by many other companies around the world. As of 2021, there are at least 15 companies operating in this space (CE Delft, 2021a) with some estimating this number to be as high as 60 (Corbyn, 2020). The American start-up Eat Just is the first to make lab-grown meat as chicken nuggets available to consumers in Singapore (Tan, 2020; Carrington, 2021; Smith & Shah, 2021). Future Meat Technologies is already producing chicken breasts at $3.90 per piece (Future Meat, 2021). According to ATKearney (2019) and based on expert interviews, cultured meat will significantly disrupt the agricultural and food industry and by 2040 most meat will not come from slaughtered animals. The experts predict that 60% of the consumed meat will either be grown in the lab or replaced by plant-based products (ATKearney, 2019).

The destiny of cultured meat will ultimately be decided by consumers and their acceptance, or lack thereof, will vet the future of mass commercialisation and production. Whether the public will accept unreservedly cultured meat is yet to be seen as its unnaturalness and artificialness may continue to provoke a sense of disgust and rejection in some consumers. It will certainly not appeal to people who prefer fruits, vegetables and other plants because of health and nutritional reasons. The progress that is happening in the field of alternative proteins is not only responding to the current planetary emergency but is likely to provide options for all.

Plant-Based Alternatives

Plant-based alternatives (also referred to as plant-based analogues or substitutes or fake meat) are seen as the new meat. They look like livestock-based meat, are designed to bleed like meat, sizzle and cook like meat, taste like meat, without an animal being involved in the whole process. Made using protein extracted from plants such as soy, pea or wheat, they are designed into familiar shapes and textures, including burgers, sausages, mince, bacon and steaks. They are perceived to be better for the planet due to their much smaller environmental footprint, but there is not enough evidence yet as to what their benefits are for human health and whether they will help reduce the rates of preventable diseases. Many claim that because these meat substitutes are plant-based, they already can potentially contribute towards lowering of cholesterol and blood sugar levels as well as reducing saturated fat and increasing fibre intake (Before the Butcher, 2020; Unilever Food Solutions, 2021).

However, there is a lack of rigorously designed independent studies that can provide convincing evidence and answer any concerns (Hu et al., 2019). For example, it is not yet known what are the long-term effects of the soy leghemoglobin – the heme used in some of the plant-based burgers. Soy and wheat are also common allergens which some people need to avoid. The use of GMOs and glyphosate in the production of soybeans is another area of concern. There are many longitudinal epidemiological studies, randomised clinical trials and short-term interventions which show that replacing animal-based meat with vegetables, fruits, nuts, seeds, legumes and other plant-based proteins is beneficial for human health. However, the evidence about the plant-based meat alternatives is yet to emerge. The fact that all these products use purified plant protein (e.g. soy protein isolate and concentrate) rather than whole foods raises the need to further understand the implications for consumers.

As of 2020, 22 companies in Australia produce plant-based alternatives (Food Frontier, 2021) – up from 14 in 2019, and with a double-digit growth in consumer interest in supermarkets in 2019, be it from a low base (Dingley, 2020). The two most popular alternatives are Beyond Burger and Impossible Burger which are being sold in tens of thousands of outlets across Australia. It is interesting that 95% of those who buy the Impossible Burger are meat-eaters (Dingley, 2020) indicating that the plant-based alternatives are contributing to a food transition towards decreased consumption of livestock-sourced products. In fact, 42% of Australians are already reducing their meat intake (Food Frontier, 2019), and from a high base, and the availability of such products is contributing to this trend. Furthermore, the plant-based alternatives have doubled the employment opportunities and manufacturing in 2020 with projections for future soaring demand domestically and globally (Food Frontier, 2021).

Plant-based alternatives are also booming in Europe which accounts for around 40% of the global market and is forecast for a further strong growth of 60% from €1.5bn in 2018 to €2.4bn by 2025 (Deloitte, 2019). The UK, in particular, is leading the way in plant-based alternatives. Despite the continuous emergence of many new and different plant-based alternative products on the existing food market, the whole industry globally is still in a relatively early lifecycle stage. It offers significant opportunity and scope for further growth and development. The global industry is foreseen to be intensified by new product development and growing consumer demand.

In the development of this novel protein arena, it is becoming increasingly important to have a tailored and innovative approach for marketing and differentiation. For example, in the food and beverage sector, the direct-to-consumer model allows producers to gain valuable insights into consumer behaviour and what drives purchasing decisions. The use of social media and other communication channels is also essential. In the plant-based alternatives sector, the direct-to-consumer model was applied by many companies, such as Allplants in UK – a meal-kit delivery service of healthy, plant-based, frozen and ready to eat meals, which raised £7.5 m in venture capital funding in September 2018 (Deloitte, 2019). Similarly, supported and promoted by the tennis star Serena Williams and the snowboarder Shaun White, a company named Daily Harvest – a plant-based vegan frozen fast-food company in USA, provides a build-your-box service delivering customer orders to their doorsteps (Deloitte, 2019). Another US-based company, a manufacturer of vegan and vegetarian foods, Amy's Kitchen launched "Amy's Drive-Thru", the first plant-based fast-food restaurant (Deloitte, 2019).

The plant-based alternatives can be seen as a stepping stone in the food transition humanity needs to do towards healthier diets and protecting the ability of the planet to sustain food production into the future. While there are no doubts that whole plant-based foods are the best option, innovation and further research in the area of plant-based alternatives can deliver some very promising results.

Algae

As another alternative protein, algae are positioned to provide future opportunities to make the food sector more sustainable and more productive. Algae are protein rich and can be grown in tanks, using just water and sunlight. Thus, they offer a great potential for formulation and preparation of low-cost, vegetarian meat alternatives (Mistry et al., 2020).

A few companies on the market are already utilising this micro-algae potential using extruded for texture Chlorella flour. An example is the Singaporean-based Sophie's Bionutrients which creates a low-sodium burger (as well as a milk alternative) in a circular bioeconomy way transforming waste from tofu, brewer spent grain and molasses from sugar refineries into novel algae-based foods (Cornall, 2021). Another company is Dambert which uses the blue-green alga Spirullina powder enriched with Omega 6. Considered a food of the future, Spirulina has been used for centuries as a source containing many beneficial micro- and macronutrients, such as high 60–70% complete protein, vitamins, gamma-linolenic acid, phycocyanin and sulphated polysaccharides (Liestianty et al., 2019). Algae protein is also successfully exploited for making high-protein pasta. Some market players in this space are Alver (2021), a company using golden Chlorella alga, and Nestlé working in partnership with Corbion using algal fermentation to create next-generation food through improved functionality, taste and nutritional profile of microalgae-based ingredients for plant-based products (Nestlé, 2019). Another market actor, Back of the Yard Algae Sciences (BYAS, 2021), is similarly finding new uses of Chlorella and Spirulina algae as sustainable, natural ingredients to address the global challenges of nutritious, affordable, with reduced environmental and social cost plant-based proteins. This also helps the need to make vertical agriculture sufficiently economically and environmentally sustainable to have a positive impact on urban food chains, and helps to transform cellular agriculture from a wasteful, early innovation into a viable source of protein for billions of people. In Japan, native sea grapes or marine algae of the genus Caulerpa (Caulerpa lentillifera, Caulerpa racemosa) known as green caviar and gourmet food are promoted as a delicacy but are yet to enter the western market (Mistry et al., 2020).

The algae food-tech landscape is growing rapidly with companies working on consumer goods and ingredients on macro (Seakyra, Algalimento, Earthrise, Ikea, New Wave, Sea Farms, Algama, Akua etc.) and micro (Algawise, Simpliigood, Olyck, Phyco Health, Alga Energy, Alga Spring, Veramaris, Algae Bulgaria, Ecoduna etc.) levels. Algae are also very versatile and a promising area for agriculture (BioAtlantis, NATRAkelp, NeoAlgae, Monzonbiotech, Algarithm, Kelpgrow, Cargill and Monsanto), chemical alternatives (Odontella, Syntheseas and SAMS), pigments (Naturex, Earthrise, Mars and NostAlgae), materials and packaging (Algotek and Checkerspot) as well as for environmental remediation (Pacific Bio, Neste, Aqualia, AlgaEnergy and Microalgal Services). Future food replacements, alternative protein, food components, fermented products, functional foods and colourants can all be derived from algae.

Found all over the Earth (Ścieszka & Klewicka, 2019), algae can also be grown artificially. Continuous exploring and future developments of algae-based applications could transform the food and beverage sectors, both from the perspective of novel products as well as the sustainability of the existing food production systems. Further algae can address agricultural waste, mitigate greenhouse gases linked to food and beverage production and can integrate waste streams into the circular bioeconomy.

Dairy-Free Protein

With the changing consumer tastes due to animal-welfare and sustainability trends, dairy-free alternatives are also establishing a presence and are projected to have a substantial market share (Saraco, 2020). The plant-based (or vegan) milks are already broadly available on the market (Marinova & Bogueva, 2020). Like tofu, soy milk in China has been popular for two millennia. It is the penetration of dairy milk alternatives in the western market that is attracting commercial attention. In USA, 41% of households already buy plant-based milks and the sector has surpassed $2 billion (Chiorando, 2020a) while in UK this share is 32% (Chiorando, 2020b).

Originating back in the sixteenth century from China, dairy-free cheese was made with fermented tofu or whole soy, and was claimed as a food high in protein, spreadable and with a strong attractive smell (Kanner, 2019). This food however is different from the western concept of cheese. It is only relatively recently that the first commercially produced cheeses with properties similar to the dairy cheeses, started to be made from soy, coconut oil, potato starch, cashews, peanuts, almonds, sesame, sunflower seeds and oat-based formulations. Currently there are more than 110 dairy-free imitation cheeses, most of which (74%) contain coconut oil as their primary ingredient (Saraco & Blaxland, 2020).

Although more efforts are needed to accurately mimic the organoleptic attributes of dairy cheeses, most plant-based cheeses are low on saturated fat and cholesterol. In 2020, the market was worth $1.15 billion and is expected to grow at a compound annual growth rate of 12.8% between 2020 and 2027 (Grand View Research, 2020). With dairy cheese being one of the food products with the highest carbon footprint, land requirements, eutrophying emissions and freshwater withdrawal, a shift to plant-based options is highly beneficial in the planetary emergency.

Artificial Fish

The future will also offer us artificial fish. If we master the cultivation of meat in a laboratory, nothing can stop us from applying the same knowledge and skills to growing fish in such conditions. All other alternative protein approaches can similarly be applied. An American company already makes synthetic shrimps from red algae (Monterey Bay Seaweeds, 2019), there are plant-based alternatives to tuna (Unilever Food Solutions, 2021; Finless Foods, 2021), and a Singaporean company is expecting to have crustaceans, such as shrimps, crabs and lobsters available commercially using cellular agriculture technology by 2022 (Shiok meats, n.d.).

3D Food

The bioprinting of synthetic meat or meat substitutes is another contemporary trend and a proposed alternative to animal-based food production. In reality, 3D food printing is in its infancy and although at the moment it consists simply of edible materials that can go through a nozzle into making intricate shapes, it can potentially help in reducing food waste and producing very personalised meals, including in unconventional cooking places, such as in space (Carolo, 2021). As with any meal, the ingredients will determine its nutritional values and environmental impacts. Artificial intelligence and 3D food printing are already utilised to produce plant-based alternatives resembling closely the molecular structure, taste and texture of meat products (Schmidinger et al., 2018).

Writing about alternative proteins, we should recall one of the best dietary advices in the words of Michael Pollan (2007, para. 1): "Eat food. Not too much. Mostly plants". Now that we have gone through all amazing developments happening in the area of alternative proteins, we need to also understand the pros and cons attached to them.

Alternative Proteins Pros

The overview of the alternative proteins focussed particularly on the promises they create with the vision for a paradigmatic shift in protein production and consumption and the ultimate potential for better future food systems (Sexton et al., 2019). Seen as technological innovations, the alternative proteins are being developed with the intention to offer many environmental benefits through reduction in greenhouse gas emissions, freshwater withdrawal, freeing hectares of land (McClements et al., 2021) as well as health, nutrition, food safety and ethical gains, particularly related to animal welfare.

These good intentions, even if they represent the state of play, need to be communicated and marketed carefully to consumers, given the fact that these are new options whose acceptance may not be straightforward. This is particularly the case with insects (van Huis, 2017) and cultured meat (Bogueva & Marinova, 2020) where clear barriers exist. Brands of alternative proteins are continuously working on improving their messaging and expanding their market shares (Ace Metrix, 2020) with some reporting that 97% of consumers who have tried the new products intend to purchase them again (Stanley, 2020). Various tactics are employed, often building on conventional meat marketing techniques, such as using celebrities, digital platforms or leveraging partnerships with established fast-food restaurant chains and supermarkets (Sweeney, 2021). The need for future research, development and commercialisation of new products has created a prospective marketplace to attract investors (Mouat & Prince, 2018), especially securing backing by rich and powerful innovators, such as Bill Gates, Richard Branson and Sergei Brin (Sexton et al.,

2019). Progressive food and other industry players, such as Tyson Foods, Cargill, Unilever, Tesco, Coles and Woolworths are also joining the race for alternative proteins (Tyson Foods, 2019; Food Frontier, 2021). Consumers so far are responding well with successful sell-outs of plant-based products in some of the world's largest food chains, such as Subway, McDonalds, Starbucks, Burger King/Hungry Jack and KFC (Food Frontier, 2021; Rivera, 2021).

Some of the alternative protein products may appeal to particular niche markets. For example, insect-based bars that are high in protein may be targeted at those adhering to a Paleo diet, or be part of the nutricentric trend which defines the relationship between food and bodily health (Scrinis, 2008), including the link between physical strength and pleasing body aesthetics. The differentiation of alternative proteins from livestock production which is viewed by some as cruel, involving animal exploitation, suffering and slaughtering, can be attractive on ethical grounds as pain-free options, irrespective of their environmental credentials.

The World Economic Forum (2019, p. 16) identifies six supportive narratives about alternative proteins:

- such foods help you live a healthier life – they are high in protein, contain more fibre, potentially avoid saturated fat components, are free from hormones and antibiotics reducing the risks of antimicrobial resistance (Van Boeckel et al., 2015; Review of Antimicrobial Resistance, 2015);
- they are free of the risk of food poisoning and contamination – conventional livestock-based meat is exposed to pathogens because of the caging, crating and crowding of animals (Haspel, 2018) which weaken their immune systems and are making them more susceptible to diseases. There are no problems with animal waste and affluents from slaughter houses. Also, handling plant-based foods involves fewer opportunities for spoilage and contamination;
- these products taste nice – this is very important for consumers and there is already evidence that they would return to these "guilt-free" options if their experience is positive (Stanley, 2020). Such narrative is particularly appealing to flexitarians, vegetarians, vegans and those who want to reduce their animal meat intake, including many Millennials and Generation Z young people;
- the environmental footprint of these food products is better – these products overall have a much smaller environmental footprint compared to livestock-based foods, particularly if renewable energy sources are potentially used for cultured meat. They will also appeal to those interested in planetary health (Marinova & Bogueva, 2019), reducing the greenhouse gas emissions of food (Herrero et al., 2011), engaged in climate change advocacy and movements (Veron, 2016), such as climate strikes and Extinction Rebellion;
- they do not harm animals – more than 80 billion animals are slaughtered for food each year; alternative proteins provide different pathways to kinder food systems which do not harm or exploit other sentient beings (Singer, 2002); animal rights have long been a sensitive and politically charged area (Garner, 2003, 2010);
- alternative proteins improve food security by releasing land currently used for grazing or production of animal feed – grains used as feed can be directed

immediately for human consumption; this is a more moral option (Raphaely & Marinova, 2014). Although not all consumers will react to such a global perspective, when combined with implications for soil fertility, biodiversity loss and climate change, intra- and intergenerational justice in relation to food security, this narrative is easier to frame.

While alternative proteins are emerging as a viable solution for humanity, there are still many unknowns. It is important to keep things into perspective to avoid another trajectory of dependence similar to fossil fuels or livestock-based foods.

Alternative Proteins Cons

Most alternative proteins involve some kind of processing and many consumers are concerned that they are not healthy. Before making a judgement, it is essential to understand what the dangers of highly-processed foods are. They are mainly linked to such foods being high in calories and having a high content of sugar, sodium or fat contributing to obesity and related non-communicable diseases. Direct consumption of whole foods, such as vegetables, fruits, tubers, legumes, wholegrains, seeds and nuts, is obviously very beneficial for human health. If people are consuming excessively animal-based foods and other highly processed items, from a health perspective they will be better off replacing their current choices with the plant-based alternative proteins.

The same publication by the World Economic Forum (2019, p. 16) identifies four cautionary narratives about alternative proteins:

- they will always play a minor role in the food system – it is easy to dismiss a new trend until a momentum develops; however, there is increasing evidence that these new products are disrupting the food system (ATKearney, 2019) and similar trends are observed with renewable energy, including battery storage, electric vehicles and blockchain technologies. As the analysis by Garnett et al. (2017) shows, it is likely that the roles will be reversed in the future and livestock will play only a minor role in the food system, mainly in places where this is the only viable option (White & Hall, 2017). This is a logical response to the planetary emergency;
- these products are not real food – this criticism is based on the fact that the alternative proteins are processed foods and may even be seen as junk food (Hoffman, 2021). Many contain soy protein concentrate with levels of nitrites comparable to bacon and are associated with increased risk of colorectal cancer – one of the reasons why the World Health Organisation (WHO) classified processed meats as Category 1 carcinogenic. Before we fall into another food trap and while waiting for research evidence to emerge about the safety or contrarywise of all alternative proteins, it may be worth remembering that there are many plant-based

safe options, such as tofu, tempeh, chickpeas, beans, lentils, vegetables and fruits that we can use to achieve better planetary health;

- alternative proteins are not as good as the real food – apart from the cost (Sexton et al., 2019), many consumers are picky about the nature of their food and often reclaim the whole foods from the ultra-processed which are made tasty and habit-forming in order to generate company profits (Monteiro, 2009); high levels of protein make the food less palatable and more additives are added to improve its taste (McClements, 2019). Synthetic colours may also be added to improve the optical appeal of the food (McClements, 2019). Under the banner of environmental benefits, many new alternative protein options may escape a thorough scrutiny by consumers, others may be rejected because of scepticism, lack of transparency and not enough information about the production process (Bogueva & Marinova, 2020). There is also distrust in the technologies used, such as lab-grown, extrusion, biomedical techniques and genetic engineering;

- livestock is more than food – livestock indeed represents livelihood and companionship to many people around the world. The livestock industry is a big contributor to national gross domestic product (GDP) and trade between nations. It is linked to tradition and also offers employment through the entire supply chain, including slaughterhouses, processing and retail facilities. We have witnessed similar arguments in relation to other industry sectors; there are, for example, parallels with the fossil-fuel-based energy sector. Another parallel that we often draw is with the tobacco industry which is still thriving in the southern parts of our native Bulgaria. It offers livelihood, employment and contributes significantly to the country's economy. The tobacco leaf was described as the Bulgarian gold and in the 1960s the country was the world's biggest exporter of cigarettes (Neuburger, 2013). Tobacco was a national pride and a cultural identity for Bulgaria. Nowadays, 35% of Bulgarians are smokers and the country is among the most affected in Europe by smoking-related diseases and mortality, with women, young people and people from socially vulnerable and disadvantaged groups most affected (Xinhua, 2019). We somehow doubt that many would vehemently approve a continuous support for the tobacco industry given all medical evidence we now have. Moreover, there are already many job and trade opportunities in the alternative protein industries.

What is then the future for alternative proteins? As the psychologist Paul Rozin (1999, p. 9) wrote, "food is fundamental, fun, frightening and far-reaching". How far will the alternative proteins reach?

The Future of Alternative Proteins

The Economist announced 2019 the year of the vegan. Plant-based food is the fastest growing food trend globally and although it could be a vehicle for more sustainable future, it is only in its infancy. Across the world, there are big increases in the number of people who declare themselves as eating only plant-based foods, be it from a very low base (Oberst, 2018). As of 2021, 14% of the world population is vegan, vegetarian or in any of the related categories (Meyer, 2021). If there is a food transition happening, are alternative proteins part of it?

The food industry is grappling with the greatest challenges since industrialisation – climate change, decreasing soil fertility and increasing population numbers. We are witnessing an unparalleled acceleration and vast advancement in science and research, including ground-breaking technological solutions employed to produce alternatives, although not always entirely new, but positioned to be one of the top future technology trends (Apostolidis, 2019). Today, the players in the alternative proteins industry believe that the future of their new plant-sourced meats, algae and insects-based products will be bright as they hinge on the taste, texture, appearance, product accessibility and affordability, upcycling waste (e.g., with insects as feed) as well as the increasing consumer perceptions about health and sustainability (Food Frontiers, 2019; Eagle, 2019).

There is no doubt that the alternative proteins, and especially the growth of the novel plant-based alternatives across the many categories, are generating an unparalleled level of competition and disruption to the current global meat and dairy sector. These alternatives are no longer for niche consumers and warrant visible shelf space in the supermarkets. The market surge for plant-based alternatives is expected to continue across all geographies (Deloitte, 2019). Cultured meat, edible insects, and algae-based alternatives similarly show high potential and together with the plant-based meat replacements are likely to disrupt the multi-billion-dollar global meat industry, with alternative proteins projected to take 60% of the market by 2040 (ATKearney, 2019).

The Economist believes that health, climate care and less animal suffering, are behind the trend and interest in alternative proteins (The Economist, 2020). However, for this to happen, many current challenges need to be overcome. They are both on the production and consumption side with some similarities in interest but different responses. There are also the challenges related to the newness of these products and lack of reliable science-based evidence about their long-term impacts.

Production Challenges

On the producer side, new or improved technologies are needed to achieve positive gastronomic experiences for the consumers. For example, the experience of eating a steak or grilled piece of meat is yet to be replicated. The umami flavour, taste, texture and appearance of plant-based alternatives need to remain unchanged even if used as leftovers the following day. For cheese products, the stringy melting sensation is also expected for ready-meal options (e.g. frozen lasagne or cheese burger). Companies using cell-, fermentation-based and any other new technologies will look to leverage any intellectual property to increase their commercialisation ambitions covering as many animal-based mimicking products as possible and contributing to the growth of the evolving meat alternatives market.

To stay on top of the game, many companies, including livestock-based and major food groups, such as Tyson, Smithfield, Perdue, Nestlé, Cargill, Hormel and Applegate Farms, are looking to secure access to the growing fake-meat, sustainable algae and other alternatives market (Yaffe-Bellany, 2019). Others, like Meat & Livestock Australia (MLA, 2020) are actively pursuing strategies to reduce the carbon footprint of animal meat and improve their environmental stewardship. Giving algae-based supplements to ruminant animals is seen as a promising way to reduce methane emissions (Schlossberg, 2020). Ultimately, it will be the consumer who will decide the future of alternative proteins.

Consumer Challenges

The Millennials, Generation Z and the flexitarian consumers are fuelling the gushing popularity of the plant-based dietary options across a large part of the affluent world (Deloitte, 2019, 2021). Consumers are experiencing an increasing awareness about the environmental, health and ethical impacts of producing and consuming meat and dairy-based products. The increased desire for healthy lifestyle choices plumped by the notion of wellness and well-being, has inspired and accelerated changes in the consumers' eating behaviour toward more sustainable options. Furthermore, there is significant media attention towards the new food products and that also influences the consumer. Novelty is often attractive and price is not always a barrier for the wealthy sections of society.

With the Millennials and Generation Z outnumbering the Baby Boomers, the consumer landscape is changing steering humanity into a new era of food (Askew, 2019). These consumers are challenging the food status quo and are demanding healthy, environmentally clean and transparently produced food (Bogueva & Marinova, 2020). Having the web on their fingertips, they will also be quick at reacting to any scientific evidence pro or against the alternative proteins.

Challenges for the Alternative Proteins

Despite the optimistic forecasts for a stable market growth for all alternative proteins, many things remain unclear in this ambitious journey and need to be addressed for the projections to become a reality. Scientific evidence is needed to confirm their environmental and health credentials.

For example, the fact that a product is plant-sourced does not suggest that it has the benefits of a diet based on whole plants, such as lower body mass index (BMI), lower rates of obesity and non-communicable diseases, such as type 2 diabetes and heart disease. The highly processed TVP is used as a base for the production of many plant-based alternatives. Having plants involved in the plant-based meat alternatives does not suggest that the end product has the benefits of the plants used. Despite being marketed as healthier options over meat, alternative plant-based meat products should be carefully assessed as they may not necessarily be healthy. Components, such as trans and saturated fats, methylcellulose, soy leghemoglobin, zinc gluconate and other new substances are already posing concerns (Integris Health, 2020).

Cell-based meat products are potentially facing the same health-related problems as conventional meat and in addition, they require proper regulation around food safety. This may be more complex than with GMO foods. The scientists will have to open the doors of their labs and become good communicators to build trust in the consumers. For the time being, the price remains a barrier but that may change with expanding manufacturing facilities in Singapore, USA, Israel and Europe. Insect-based products similarly require stringent regulation (FAO, 2021).

There are also socio-cultural perspectives. How will the connotation between meat and masculinity manifest around plant-based alternatives (Bogueva et al., 2022)? Will the many athletes who adhere to plant-based diets be able to influence the macho men and their desire to impress the "chicks"? What would be the halal and kosher interpretations as well as the advice from any other religions?

The environmental footprint of the alternative proteins will also need to be measured and communicated. People do not want to be susceptible to manipulation one way or another. It is clear that the food of the future will change but many options will remain the same, unless the consumer forgets about them. Indeed, in the search for novel food solutions, it would be a shame to forget about eating fruits, vegetables, grains, legumes, nuts and seeds and the myriad of beautiful plants that nature already has in store for us and that do not cost the Earth.

Concluding Thoughts

Humanity has relied on livestock animals for quite a long time and they may continue to have a place in sustainable food production systems, but a very limited place (Garnett et al., 2017). The advantages and potential of alternative proteins – some new, others used for millennia, are opening a new perspective into maintaining planetary health and responding to the current environmental and climate emergencies. While consumers are still navigating the dietary choices they make and while the jury is out about the long-term benefits of plant-sourced, insect-based or cultured meat options, the diversity, appeal and advantages of staple foods, such as fruits, vegetables, legumes, nuts, grains and seeds should not be ignored or dismissed. These foods hold many clear immediate answers for the planetary emergency.

References

Ace Metrix. (2020). *How brands are serving their plant-based "meat" ads*. https://www.acemetrix.com/insights/blog/plant-based-meat-ads/

Adamson, M. W. (2004). *Food and medieval times*. Greenwood Press.

Alam, M. S., Kaur, J., Khaira, H., & Gupta, K. (2016). Extrusion and extruded products: Changes in quality attributes as affected by extrusion process parameters: A review. *Critical Reviews in Food Science and Nutrition, 56*(3), 445–473. https://doi.org/10.1080/10408398.2013.779568

Alver. (2021). *Good for you and our planet: Enjoy delicious high protein foods for daily life with Golden Chlorella*. https://www.alver.ch

Anderson, E. N. (2014). *Food and environment in early and medieval China*. University of Pennsylvania Press.

Apostolidis, C. (2019). From "yucky" to "yummy": Drivers and barriers in the meat alternatives market. In D. Bogueva, D. Marinova, T. Raphaely, & K. Schmidinger (Eds.), *Environmental, health and business opportunities in the new meat alternatives market* (pp. 1–18). IGI Global.

Apostolidis, C., & McLeay, F. (2016). It's not vegetarian, it's meat-free! Meat eaters, meat reducers and vegetarians and the case of Quorn in the UK. *Social Business, 6*(3), 267–290. https://doi.org/10.1362/204440816X14811339560938

Askew, K. (2019). *'Food will define the future': Compass CEO sees tech and trends transforming the food system*. Food Navigator. https://www.foodnavigator.com/Article/2019/03/01/Food-will-define-the-future-Compass-CEO-sees-tech-and-trends-transforming-the-food-system

ATKearney. (2019). *How will cultured meat and alternative meat products disrupt the agricultural and food industry?* https://www.kearney.com/documents/20152/2795757/How+Will+Cultured+Meat+and+Meat+Alternatives+Disrupt+the+Agricultural+and+Food+Industry.pdf/06ec385b-63a1-71d2-c081-51c07ab88ad1?t=1559860712714

Axworthy, N. (2019). *Price of lab-grown meat to plummet from $280,000 to $10 per patty by 2021*. Veg news. https://vegnews.com/2019/7/price-of-lab-grown-meat-to-plummet-from-280000-to-10-per-patty-by-2021

Back of the Yard Algae Sciences (BYAS). (2021). *Our products*. https://www.algaesciences.com

Bashi, Z., McCullough, R., Ong, L., & Ramirez, M. (2019). *Alternative proteins: The race for market share is on*. McKinsey & Company. https://www.mckinsey.com/~/media/McKinsey/Industries/Agriculture/Our%20Insights/Alternative%20The%20race%20for%20market%20share%20is%20on/Alternative-proteins-The-race-for-market-share-is-on.pdf

BBC. (n.d.). *World War Two foods that changed the way we eat*. BBC Food. https://www.bbc.co.uk/food/articles/world_war_two_foods

Before the Butcher. (2020). *9 surprising benefits of plant-based meat substitutes*. https://btbfoods.com/blog/9-surprising-benefits-of-plant-based-meat-substitutes/

Beniwal, A. S., Singh, J., Kaur, L., Hardacre, A., & Singh, H. (2021). Meat analogs: Protein restructuring during thermomechanical processing. *Comprehensive Reviews in Food Science and Food Safety, 20*(2), 1221–1249. https://doi.org/10.1111/1541-4337.12721

Bhat, Z. F., Kumar, S., & Fayaz, H. (2015). In vitro meat production: Challenges and benefits over conventional meat production. *Journal of Integrative Agriculture, 14*(2), 241–248. https://doi.org/10.1016/S2095-3119(14)60887-X

Bogueva, D., & Marinova, D. (2020). Cultured meat and Australia's Generation Z. *Frontiers in Nutrition, 7*, 148. https://doi.org/10.3389/fnut.2020.00148

Bogueva, D., Marinova, D., & Phau, I. (2018). Is meat a luxury? In D. Bogueva, D. Marinova, & T. Raphaely (Eds.), *Handbook of research on social marketing and its influence on animal origin food product consumption* (pp. 172–186). IGI Global.

Bogueva, D., Marinova, D., & Bryant, C. (2022). Meat me halfway: Sydney meat-loving men's restaurant experience with alternative plant-based proteins. *Sustainability, 14*(3), 1290. https://doi.org/10.3390/su14031290

Bramen, L. (2010). *Seitan: The other fake meat*. Smithsonian Magazine. https://www.smithsonianmag.com/arts-culture/seitan-the-other-fake-meat-97622092/

Brennan, T., Katz, J., Quint, Y., & Spencer, B. (2021). *Cultivated meat: Out of the lab, into the frying pan*. McKinsey & Company. https://www.mckinsey.com/industries/agriculture/our-insights/cultivated-meat-out-of-the-lab-into-the-frying-pan

Carolo, L. (2021). *3D printed food: All you need to know in 2021*. All3DP. https://all3dp.com/2/3d-printed-food-3d-printing-food/

Carrington, D. (2021). *No-kill, lab-grown meat to go on sale for first time*. The Guardian. https://www.theguardian.com/environment/2020/dec/02/no-kill-lab-grown-meat-to-go-on-sale-for-first-time

CE Delft. (2021a). *LCA of cultivated meat: Future projections of different scenarios*. https://cedelft.eu/wp-content/uploads/sites/2/2021/04/CE_Delft_190107_LCA_of_cultivated_meat_Def.pdf

CE Delft. (2021b). *TEA of cultivated meat: Future projections of different scenarios*. https://cedelft.eu/wp-content/uploads/sites/2/2021/04/CE_Delft_190254_TEA_of_Cultivated_Meat_Def.pdf

Chapman, H. A., & Anderson, A. K. (2012). Understanding disgust. *Annals of the New York Academy of Sciences, 1251*(1), 62–76. https://doi.org/10.1111/j.1749-6632.2011.06369.x

Chen, X., Feng, Y., & Chen, Z. (2009). Common edible insects and their utilization in China. *Entomological Research, 39*, 299–303. https://doi.org/10.1111/j.1748-5967.2009.00237.x

Chiorando, M. (2020a). *41% of US households now buy vegan milk, says data*. Plant-based News. https://plantbasednews.org/news/41-us-households-buy-vegan-milk/

Chiorando, M. (2020b). *A third of UK households now buy dairy-free milk*. Plant-based News. https://plantbasednews.org/lifestyle/32-u-k-households-buy-dairy-free-milk/

Chriki, S., & Hocquette, J.-F. (2020). The myth of cultured meat: A review. *Frontiers in Nutrition, 7*, 7. https://doi.org/10.3389/fnut.2020.00007

Churchill, W. (1931). *Fifty years hence*. https://teachingamericanhistory.org/library/document/fifty-years-hence/

Collingham, L. (2011). *The taste of war: World War II and the battle for food*. Penguin Books.

Corbyn, Z. (2020). *Out of the lab and into your frying pan: The advance of cultured meat*. The Guardian. https://www.theguardian.com/food/2020/jan/19/cultured-meat-on-its-way-to-a-table-near-you-cultivated-cells-farming-society-ethics

Cornall, J. (2021). *Sophie's Bionutrients develops world's first dairy-free micro-algae based milk alternative*. Dairy Reporter. https://www.dairyreporter.com/Article/2021/05/04/Sophie-s-Bionutrients-develops-world-s-first-dairy-free-micro-algae-based-milk-alternative#

Deloitte. (2019). *Plant-based alternatives: Driving industry M&A*. https://www2.deloitte.com/content/dam/Deloitte/uk/Documents/consumer-business/deloitte-uk-plant-based-alternatives.pdf

Deloitte. (2021). *A call for accountability and action: The Deloitte Global 2021 Millennials and Gen Z survey*. https://www2.deloitte.com/global/en/pages/about-deloitte/articles/millennial-survey.html

Dictionary.com. (n.d.). Disgust. In *Dictionary.com*. Retrieved July 18, 2021, from https://www.dictionary.com/browse/disgusted

Dingley, M. (2020). *Your guide to fake meat in Australia in 2020*. Matthews Intelligent Identification. https://www.matthews.com.au/blog/your-guide-to-fake-meat-in-australia-in-2020

Du Bois, C. M. (2018). *The story of soy*. Reaktion Books.

Eagle, J. (2019). *Microalgae protein grown in tanks to be the next generation future of food*. Food Navigator. https://www.foodnavigator.com/Article/2019/02/13/Microalgae-protein-grown-in-tanks-to-be-the-next-generation-future-of-food

Egolf, A., Siegrist, M., & Hartmann, C. (2018). How people's food disgust sensitivity shapes their eating and food behaviour. *Appetite, 127*, 28–36. https://doi.org/10.1016/j.appet.2018.04.014

Finless Foods. (2021). *Plant-based and cell-cultured seafood that changes how the world eats... and supports a thriving ocean*. https://finlessfoods.com/

Finnigan, T., Wall, B. T., Wilde, P. J., Stephens, F. B., Taylor, S. L., & Freedman, M. R. (2019). Mycoprotein: The future of nutritious nonmeat protein, a symposium review. *Current Developments in Nutrition, 3*(6), nzz021. https://doi.org/10.1093/cdn/nzz021

Food and Agriculture Organisation of the United Nations (FAO). (2018). *Meat products with high levels of extenders and fillers*. https://documents.pub/download/meat-products-with-high-levels-of-extenders-and-fillers

Food and Agriculture Organisation of the United Nations (FAO). (2021). *Looking at edible insects from a food safety perspective. Challenges and opportunities for the sector.* https://doi.org/10.4060/cb4094en

Food Frontier. (2019). *Hungry for plant-based: Australian consumer insights.* https://www.foodfrontier.org/wp-content/uploads/2019/10/Hungry-For-Plant-Based-Australian-Consumer-Insights-Oct-2019.pdf#gf_2

Food Frontier. (2021). *2020 state of the industry: Australia's plant-based meat sector.* https://www.foodfrontier.org/wp-content/uploads/dlm_uploads/2021/03/Food-Frontier-2020-State-of-the-Industry.pdf

Future Meat. (2021). *Bringing cultured meat to the table: Delicious.* Healthy. Sustainable. https://future-meat.com

Garner, R. (2003). Political ideologies and the moral status of animals. *Journal of Political Ideologies, 8*(2), 233–246. https://doi.org/10.1080/13569310306087

Garner, R. (2010). Animal rights, political theory and the liberal tradition. *Contemporary Politics, 8*(1), 7–22. https://doi.org/10.1080/13569770220130095

Garnett, T., Godde, C., Muller, A., Röös, E., Smith, P., de Boer, I. J. M., Zu Ermgassen, E., Herrero, M., van Middelaar, C. E., Schader, C., & van Zanten, H. H. E. (2017). *Grazed and confused? Ruminating on cattle, grazing systems, methane, nitrous oxide, the soil carbon sequestration question – And what it all means for greenhouse gas emissions.* Food Climate Research Network. https://www.oxfordmartin.ox.ac.uk/downloads/reports/fcrn_gnc_report.pdf

Grand View Research. (2020). *Vegan cheese market size, share & trends analysis report by product (mozzarella, cheddar, parmesan, ricotta, cream), by source (cashew, soy), by end-use (household, foodservice), by region, and segment forecasts, 2020–2027.* https://www.grandviewresearch.com/industry-analysis/vegan-cheese-market

Groves, M. (2018). *Vegan-meat-substitutes: The ultimate guide.* Healthline. https://www.healthline.com/nutrition/vegan-meat-substitutes

Guo, X., Shao, X., Trishna, S., Marinova, D., & Hossain, A. (2019). Soybeans consumption and production in China: Sustainability perspective. In D. Bogueva, D. Marinova, T. Raphaely, & K. Schmidinger (Eds.), *Environmental, health and business opportunities in the new meat alternatives market* (pp. 124–142). IGI Global.

Hachmeister, K. A., & Fung, D. Y. (1993). Tempeh: A mold-modified indigenous fermented food made from soybeans and/or cereal grains. *Critical Reviews in Microbiology, 19*(3), 137–188. https://doi.org/10.3109/10408419309113527

Hartmann, C., & Siegrist, M. (2020). Disgust and eating behaviour. In H. L. Meiselman (Ed.), *Handbook of eating and drinking: Interdisciplinary perspectives* (pp. 315–332). Springer.

Haspel, T. (2018). *Lab-grown meat and the fight over what it can be called, explained.* Vox. https://www.vox.com/2018/8/30/17799874/lab-grown-meat-memphis-just-animal-cell

Herrero, M., Gerber, P., Vellingac, T., Garnett, T., Leipe, A., Opio, C., Westhoekf, H. J., Thorntona, P. K., Oleseng, J., Hutchings, N., Montgomery, H., Soussanai, J.-F., Steinfeld, H., & McAllister, T. A. (2011). Livestock and greenhouse gas emissions: The importance of getting the numbers right. *Animal Feed Science and Technology, 166–167*(2011), 779–782. https://doi.org/10.1016/j.anifeedsci.2011.04.083

Hexa Research. (2019). *Global tempeh market worth USD258.7 million by 2025: Hexa Research.* PR News wire. https://www.prnewswire.com/news-releases/global-tempeh-market-worth-usd-258-7-million-by-2025-hexa-research-300814521.html

Hoffman, R. (2021). *Plant-based burgers: Should some be considered 'junk food'?* The Conversation. https://theconversation.com/plant-based-burgers-should-some-be-considered-junk-food-163514

Hu, F. B., Otis, B. O., & McCarthy, G. (2019). Can plant-based meat alternatives be part of a healthy and sustainable diet? *Journal of the American Medical Association (JAMA), 322*(16), 1547–1548. https://doi.org/10.1001/jama.2019.13187

Integris Health. (2020). *Are plant based meats healthier than regular meat.* https://integrisok.com/resources/on-your-health/2020/october/are-plant-based-meats-healthier-than-regular-meat#:~:text=As%20far%20as%20sodium%2C%20calories,cancer%20and%20 type%2D2%20diabetes

Kanner, E. (2019). *The cult and culture of vegan cheese*. Holistic Primary Care. https://holisticprimarycare.net/topics/cooking-for-health/the-cult-and-culture-of-vegan-cheese/

Lager, M. M. (1945). *The useful soybean: A plus factor in modern living*. McGraw-Hill Book Company.

Lane, M. (2011). *Disgust: How did the word change so completely?* BBC News Magazine. https://www.bbc.com/news/magazine-15619543

Lawrence, S., & King, T. (2019). *Meat the alternative: Australia's $3 billion dollar opportunity*. Food Frontier. https://www.foodfrontier.org/wp-content/uploads/2019/09/Meat-the-Alternative-Food-Frontier.pdf

Liestianty, D., Rodianawati, I., Arfah, R. A., Assa, A., Patimah, S., & Muliadi. (2019). Nutritional analysis of spirulina sp to promote as superfood candidate. *IOP Conference Series: Material Science and Engineering, 509*, 012031. https://doi.org/10.1088/1757-899X/509/1/012031

Lindsay, J. (2018). *What is Quorn made of?* Metro. https://metro.co.uk/2018/03/04/what-is-quorn-made-of-7359772/

Liu, K. S., & Hsieh, F.-H. (2008). Protein–protein interactions during high-moisture extrusion for fibrous meat analogues and comparison of protein solubility methods using different solvent systems. *Journal of Agricultural and Food Chemistry, 56*, 2681–2687. https://doi.org/10.1021/jf073343q

MacEvilly, C. (2000). Bugs in the system. *Nutrition Bulletin, 25*(4), 267–268. https://doi.org/10.1046/j.1467-3010.2000.00068.x

Marinova, D., & Bogueva, D. (2019). Planetary health and reduction in meat consumption. *Sustainable Earth, 2*(3). https://doi.org/10.1186/s42055-019-0010-0. or https://rdcu.be/b2Riz

Marinova, D., & Bogueva, D. (2020). *Which 'milk' is best for the environment? We compared dairy, nut, soy, hemp and grain milks*. The Conversation. https://theconversation.com/which-milk-is-best-for-the-environment-we-compared-dairy-nut-soy-hemp-and-grain-milks-147660

Marshall, P., & Marinova, D. (2019). Health benefits of eating more plant foods and less meat. In D. Bogueva, D. Marinova, T. Raphaely, & K. Schmidinger (Eds.), *Environmental, health and business opportunities in the new meat alternatives market* (pp. 38–61). IGI Global.

McClements, D. J. (2019). *Future foods: How modern science is transforming the way we eat*. Springer.

McClements, D. J., Barrangou, R., Hill, C., Kokini, J. L., Lila, M. A., Meyer, A. S., & Yu, L. (2021). Building a resilient, sustainable, and healthier food supply through innovation and technology. *Annual Review of Food Science and Technology, 12*(1), 1–28. https://doi.org/10.1146/annurev-food-092220-030824

McGuiness, L. (2019). Essential summer seitan. *Vegetarian Journal, 38*(2), 6–11. https://www.vrg.org/journal/vj2019issue2/VJ_issue2_2019.pdf

Meat & Livestock Australia (MLA). (2020). *Stop the denigration – Time for a truthful conversation*. https://www.mla.com.au/news-and-events/industry-news/stop-the-denigration%2D%2Dtime-for-a-truthful-conversation/#

Merriam-Webster. (n.d.). Make someone's gorge rise. In *Merriam-Webster.com dictionary*. Retrieved July 18, 2021, from https://www.merriam-webster.com/dictionary/make%20someone%27s%20gorge%20rise

Meticulous Research. (2019). *Edible Insects market by product type (whole insect, insect powder, insect meal), insect type (crickets, black soldier fly, mealworms), application (animal feed, protein bar and shakes, bakery, confectionery, beverages): Global forecast to 2027*. https://www.meticulousresearch.com/product/edible-insects-market-5156

Meyer, M. (2021). *This is how many vegans are in the world right now (2021 update)*. The You. https://thevou.com/lifestyle/2019-the-world-of-vegan-but-how-many-vegans-are-in-the-world/

Miller, S. B. (2014). *Disgust: The gatekeeper emotion*. Routledge.

Mintel. (2013). *Meat-free and free-from foods – UK*. Mintel International Group Limited. https://store.mintel.com/report/meat-free-and-free-from-foods-uk-september-2013

Mistry, M., George, A., & Thomas, S. (2020). Alternatives to meat for halting the stable to table continuum – An update. *Arab Journal of Basic and Applied Sciences, 27*(1), 324–334. https://doi.org/10.1080/25765299.2020.1807084

Monteiro, C. A. (2009). Nutrition and health. The issue is not food, nor nutrients, so much as processing. *Public Health Nutrition, 12*(5), 729–731. https://doi.org/10.1017/S1368980009005291

Monterey Bay Seaweeds. (2019). *Red algae: New artificial shrimp are made from algae.* http://www.montereybayseaweeds.com/the-seaweed-source/2019/1/11/new-artificial-shrimp-are-made-from-algae

Mordor Intelligence. (2021). *Tempeh market – Growth, trends, Covid-19 impact, and forecasts (2021–2026).* https://www.mordorintelligence.com/industry-reports/tempeh-market

Mouat, M. J., & Prince, R. (2018). Cultured meat and cowless milk: On making markets for animal-free food. *Journal of Cultural Economy, 11*(4), 315–329. https://doi.org/10.108 0/17530350.2018.1452277

Nestlé. (2019). *Nestlé to partner with Corbion for the development of microalgae-based ingredients for plant-based products.* https://www.nestle.com/randd/news/allnews/partnership-corbion-microalgae-plant-based-products

Neuburger, M. C. (2013). *Balkan smoke: Tobacco and the making of modern Bulgaria.* Cornell University Press.

Nguyen, A. (2012). *Asian tofu: Discover the best, make your own, and cook it at home.* Ten Speed Press.

Niassy, S., & Ekesi, S. (2017). *Eating insects has long made sense in Africa. The world must catch up.* The Conversation. https://theconversation.com/eating-insects-has-long-made-sense-in-africa-the-world-must-catch-up-70419

Oberst, L. (2018). *Why the global rise in vegan and plant-based eating isn't a fad (600% increase in U.S. vegans + other astounding stats).* Food Revolution Network. https://foodrevolution.org/blog/vegan-statistics-global/

Perren, R. (2006). *Taste, trade and technology: The development of the international meat industry since 1840.* Ashgate Publishing.

Pollan, M. (2007). *Unhappy meals.* The New York Times Magazine. https://www.nytimes.com/2007/01/28/magazine/28nutritionism.t.html

Post, M. J. (2012). Cultured meat from stem cells: Challenges and prospects. *Meat Science, 92,* 297–301. https://doi.org/10.1016/j.meatsci.2012.04.008

Praestiin, J. (2020). *Sanitarium Health and Wellbeing, South Pacific Division.* Encyclopedia of Seventh-Day Adventists. https://encyclopedia.adventist.org/article?id=A842#fn41

Quorn. (2021). *Is Quorn safe?* https://www.quorn.co.uk/faqs/general#IsQuornSafe

Raphaely, T., & Marinova, D. (2014). Flexitarianism: A more moral dietary option. *International Journal of Sustainable Society, 6*(1/2), 189–211. https://doi.org/10.1504/IJSSOC.2014.057846

Review of Antimicrobial Resistance. (2015). *Antimicrobials in agriculture and the environment: Reducing unnecessary use and waste.* http://bit.ly/2d36sEH

Rivera, D. (2021). *All the fast-food chains and grocers serving plant-based meat in 2021.* UpRoxx. https://uproxx.com/life/fast-food-chains-serving-plant-based-meat-2021/

Rozin, P. (1999). Food is fundamental, fun, frightening, and far-reaching. *Social Research, 66*(1), 9–30.

Rozin, P., & Fallon, A. (1987). A perspective on disgust. *Psychological Review, 94*(1), 23–41. https://doi.org/10.1037/0033-295X.94.1.23

Sanitarium. (2021). *Moments that made us.* https://www.sanitarium.com.au/about/sanitarium-story/history

Saraco, M. (2020). Functionality of the ingredients used in commercial dairy-free imitation cheese and analysis of cost-related, food safety and legal implications. *The Journal of Vegan Food, 1,* EA000001. http://www.mnsaraco.com.ar/JournalVeganFood/published/Volume_1/EA00001.php

Saraco, M., & Blaxland, J. (2020). Dairy-free imitation cheese: Is further development required? *British Food Journal, 122*(12), 3727–3740. https://doi.org/10.1108/BFJ-11-2019-0825

Schlossberg, T. (2020). *An unusual snack for cows, a powerful fix for climate: Feeding them seaweed slashes the amount of methane they burp into the atmosphere.* https://www.washingtonpost.com/climate-solutions/2020/11/27/climate-solutions-seaweed-methane/

Schmidinger, K., Bogueva, D., & Marinova, D. (2018). New meat without livestock. In D. Bogueva, D. Marinova, & T. Raphaely (Eds.), *Handbook of research on social marketing and its influence on animal origin food product consumption* (pp. 344–361). IGI Global.

Schwarz, R. W. (1970). *John Harvey Kellogg, M.D.* Southern Publishing Association.

Ścieszka, S., & Klewicka, E. (2019). Algae in food: A general review. *Critical Reviews in Food Science and Nutrition, 59*(21), 3538–3547. https://doi.org/10.1080/10408398.2018.1496319

Scrinis, G. (2008). On the ideology of nutritionism. *Gastronomica, 8*(1), 39–48. https://doi.org/10.1525/gfc.2008.8.1.39

Sexton, A. E., Garnett, T., & Lorimer, J. (2019). Framing the future of food: The contested promises of alternative proteins. *Environment and Planning E: Nature and Space, 2*(1), 47–72. https://doi.org/10.1177/2514848619827009

Shiok Meats. (n.d.). *Seafood, reinvented.* https://shiokmeats.com/

Shprintzen, A. D. (2013). *The vegetarian crusade: The rise of an American reform movement, 1817–1921.* University of North Carolina Press.

Shurtleff, W., & Aoyagi, A. (2004). *Dr. John Harvey Kellogg and Battle Creek Foods: Work with soy, history of soybeans and soyfoods, 1100 B.C. to the 1980s.* Soyinfo Center. https://www.soyinfocenter.com/HSS/john_kellogg_and_battle_creek_foods.php

Shurtleff, W., & Aoyagi, A. (2007). *History of tempeh: A special report on the history of traditional fermented soyfoods.* Soyinfo Center. http://luk.tsipil.ugm.ac.id/itd/artikel/Shurtleff-Aoyagi-HistoryOfTempeh.pdf

Shurtleff, W., & Aoyagi, A. (2013). *History of tofu and tofu products (965 CE to 2013).* Soyinfo Center. https://www.soyinfocenter.com/pdf/163/Tofu.pdf

Shurtleff, W., Aoyagi, A., & Huang, H. T. (2014). *History of soybeans and soyfoods in China and Taiwan, and in Chinese cookbooks, restaurants, and Chinese work with soyfoods outside China (1024 BCE to 2014).* Soyinfo Center. https://www.soyinfocenter.com/pdf/176/Chin.pdf

Singer, P. (2002). *Animal liberation.* Ecco.

Smith, A., & Shah, S. (2021). *The government needs an innovation policy for alternative meats.* Issues in Science and Technology. https://issues.org/alternative-meat-innovation-environment-animals/

Sogari, G., Bogueva, D., & Marinova, D. (2019). Australian consumers' response to insects as food. *Agriculture, 9*(5), 108. https://doi.org/10.3390/agriculture9050108

Spencer, C. (1993). *The heretic's feast: A history of vegetarianism.* Fourth Estate.

Stanley, T. L. (2020). *Lockdown fake meat buyers turned into full-blown plant-based converts.* Adweek. https://www.adweek.com/brand-marketing/lockdown-fake-meat-buyers-plant-based-converts/?utm_content=position_7&utm_source=postup&utm_medium=email&utm_campaign=FirstThingsFirst_Newsletter_201102054642&lyt_id=196724

Sweeney, E. (2021). *Plant-based meat brands leverage digital strategies to build brand awareness & encourage trials.* https://insights.digitalmediasolutions.com/articles/alternative-meats-digital-advertising

Tan, C. B. (2008). Tofu and related products in Chinese foodways. In C. M. Du Bois, C.-B. Tan, & S. W. Mintz (Eds.), *The world of soy* (pp. 99–120). National University of Singapore Press.

Tan, A. (2020). *Cultured meat: No-kill products may be food for the future.* The Strait Times. https://www.straitstimes.com/singapore/environment/no-kill-products-may-be-food-of-the-future

The Economist. (2020). *Interest in veganism is surging.* https://www.economist.com/graphic-detail/2020/01/29/interest-in-veganism-is-surging

Tyson Foods. (2019). *Tyson Foods unveils alternative protein products and new Raised & Rooted® Brand.* https://www.tysonfoods.com/news/news-releases/2019/6/tyson-foods-unveils-alternative-protein-products-and-new-raised-rootedr

Unilever Food Solutions. (2021). *The future of meat: The rise in plant based meat substitutes.* https://www.unileverfoodsolutions.us/chef-inspiration/plant-based-eating/trends/meat-alternatives.html

Van Boeckel, T. P., Brower, C., Gilbert, M., Grenfell, B. T., Levin, S. A., Robinson, T. P., Teillant, A., & Laxminarayan, R. (2015). Global trends in antimicrobial use in food animals. *Proceedings of the National Academy of Sciences of the United States of America (PNAS), 112*(18), 5649–5654. https://www.pnas.org/content/112/18/5649

Van Huis, A. (2017). Edible insects: Marketing the impossible? *Journal of Insects as Food and Feed, 3*(2), 67–68. https://doi.org/10.3920/JIFF2017.x003

Veron, O. (2016). (Extra)ordinary activism: Veganism and the shaping of hemeratopias. *International Journal of Sociology and Social Policy, 36*(11–12), 756–773. https://doi.org/10.1108/IJSSP-12-2015-0137

Watson, E. (2019). *Watch: Grasshoppers, not crickets, will drive the edible insect revolution, says Hargol Foodtech*. Food Navigator. https://www.foodnavigator-usa.com/Article/2019/04/12/Grasshoppers-not-crickets-will-drive-the-edible-insect-revolution-says-Hargol-Foodtech-at-Foodbytes#

Watson, E. (2021). *When will cell-cultured meat reach price parity with conventional meat?* Food Navigator. https://www.foodnavigator-usa.com/Article/2021/03/15/When-will-cell-cultured-meat-reach-price-parity-with-conventional-meat

White, R. R., & Hall, M. B. (2017). Nutritional and greenhouse gas impacts of removing animals from US agriculture. *Proceedings of the National Academy of Sciences of the United States of America (PNAS), 114*(48), E10301–E10308. https://doi.org/10.1073/pnas.1707322114

Whitnall, T., & Pitts, N. (2019). *Global trend in meat consumption*. ABARES Agricultural commodities. https://www.agriculture.gov.au/sites/default/files/sitecollectiondocuments/abares/agriculture-commodities/AgCommodities201903_MeatConsumptionOutlook_v1.0.0.pdf

Wong, A. Y.-T., & Chan, A. W.-K. (2016). Genetically modified foods in China and the United States: A primer of regulation and intellectual property protection. *Food Science and Human Wellness, 5*(3), 124–140. https://doi.org/10.1016/j.fshw.2016.03.002

World Economic Forum. (2019). *Meat: The future series*. Alternative proteins. The White Paper. http://www3.weforum.org/docs/WEF_White_Paper_Alternative_Proteins.pdf

Wynn, K., & Sebastian, B. (2019). *Growth opportunities for Australian food and agribusiness: Economic analysis and market sizing*. CSIRO Futures. https://research.csiro.au/foodag/sustainable-solutions/

Xinhua. (2019). *Bulgaria seeks to ban tobacco advertising, promotion and sponsorship*. http://www.xinhuanet.com/english/2019-09/09/c_138378813.htm

Yaffe-Bellany, D. (2019). *The new makers of plant-based meat? Big meat companies*. The New York Times. https://www.nytimes.com/2019/10/14/business/the-new-makers-of-plant-based-meat-big-meat-companies.html

Chapter 8
Food Marketing in a Planetary Emergency

Abstract Our relationship with the marketing of food occurs spontaneously, often unconsciously and although we think we have control, the choices we make are not happening in a vacuum. The chapter examines the role of food marketing highlighting the need for a change towards facilitating a sustainability transition. It elaborates on what food marketing should be delivering to the consumer in responding to the planetary emergency from a social and individual perspective. Food marketing should be supporting ethical and sustainable products with transparency about the food production process while aiming at achieving population health and wellness.

A 2019 advertisement on large billboards across Australia was saying simply "Look up". We saw it in Sydney and in Perth. In fact, 92% of the Australian people living in and around the country's capital cities saw it (Look Up, n.d.). Supported by the Outdoor Media Association, the advertising campaign claims that "great things happen to your brain & your life, when you look up & out" (Kerr & Maze, 2019, p. 8). We looked up and saw the blue sky… The advertisement was encouraging people to reconnect with nature and other human beings instead of being constantly focussed on digital devices. In that split second when we looked up, the advertisement changed our behaviour. It is more difficult to judge whether it had a longer-term effect on those who saw it. However, if in that split second you are making a decision about buying a product, the advertisers would have achieved their goal. Actually, this is not completely true as good marketing these days is not trying to just capture your short-term attention, it aims at developing a relationship between you and the brand, its products or services. Marketing uses ammunition from psychology, anthropology, cognitive neuroscience and complex systems engineering (Kerr & Maze, 2019) to create an effective call for action. Nowhere is this as true as is with the marketing of food.

Since its inception, food marketing has been aimed at following people's psychology reflecting the continuously changing consumer behaviour and taking into account various factors, including external – cultural, demographic, natural, economic, technological, political, etc., internal – personality, life stage, identity,

153

D. Marinova, D. Bogueva, *Food in a Planetary Emergency*,
https://doi.org/10.1007/978-981-16-7707-6_8

beliefs, attitudes, motivations, emotions, feelings, self-concept, knowledge, practices etc., and also organisational – brand, advertising, product, price, service, promotion, etc., factors. Food marketing also tries to capture consumers in the place where they are and offer them the most appealing information in the most effective way to influence their behaviour. It builds on key success factors, such as appealing to current trends, keeping the message simple, succinctly communicating the value proposition, personalising the call-to-act and advertising the experience, and how it will make you feel, rather than a product (Smith, 2020). The marketing of food creates impulses to self-identify with certain eating choices, to form nostalgic attachments or interdependence with foodstuff which become part of the daily routine and emotional bond – the love for a particular food. It generates an environment where consumers can easily be tempted, brought into or deluded to remain loyal caught in the clutches of advertising messages.

Our relationship with food marketing occurs spontaneously, and often unconsciously, as an engagement with or response to dominant trends. Ultimately marketing is about influencing, or even manipulating, human behaviour and it takes deep-thinking social critics to expose its effects and unravel its doings. Let's look at a few notable books. Eric Schlosser's *Fast Food Nation* (2002) was an eye-opening depiction of human consumption defined and shaped by the marketing of the fast-food industry. Marketing can be used without any regard of human health with advertising campaigns carefully tailored to appeal to specific population groups. Marion Nestle writes about this in *Food Politics* (2013), a book which exposes the efforts of lobbyist groups, political manoeuvres, public relations and aggressive advertising to lure consumers into often deceptive food choices. Michael Pollan's *The Omnivore's Dilemma* (2006) is a book which explains the paradox of food choices with marketing being responsible for America's burgeoning eating disorders, mass confusion over changing food trends with conflicting fad diets and endless supply of processed foods with improbable health claims. In *Meathooked*, Marta Zaraska (2016, p. 3) writes about people's obsession with meat and that this "international love of animal protein is not only messing up our health, it's also damaging the planet". She explains that despite the guilt people feel towards innocent farm animals, the damaged arteries caused by LDL (low-density lipoprotein or bad) cholesterol and the pollution inflicted on the planet, the human race is not giving up meat because of the hooks that are holding us up. They include genetics and culture, but also government policies and the power of the meat industry. Another addition to the book list is Carol J. Adams' *The Sexual Politics of Meat* (1990) where she exposes the patriarchal culture that is constantly finding new ways to uphold and reinforce the association between meat-eating and toxic masculinity, including the perpetuation of this myth through pervasive advertising.

As these books, reveal, our food choices are not made in a vacuum. We are put in a situation to believe that we have made an informed decision ourselves. In reality, we cannot do so, particularly if we are oblivious of the ways food companies influence our choices. These companies simply encourage consumers to eat more to help them generate sales and increase the industry's income (Nestle, 2013). Interestingly, if we want to eat healthy and sustainable food, there is no need for marketing as these choices are neither complicated, nor insipid, they are even rewarding for our bodies and the planet. We do not need to wait for new products to be created and advertised to us. There is abundance of good food choices with the broadly available fruits and vegetables, at least in the West. Developing countries also have a rich diversity of local fruits and vegetables which are increasingly being neglected in favour of unhealthier options (Ganry et al., 2011). Food marketing works to promote the social acceptability of particular foods (Nestle, 2013), which are often unhealthy, unsustainable and heavily processed. Such foods also create addicting behaviours and result in detrimental relationships that trigger obesity epidemics and high prevalence of diet-based non-communicable diseases. Rarely has food marketing been held responsible for such consequences.

From the perspective of the current environment and climate emergency, we also need to examine the role of food marketing and highlight the need for it to change and facilitate a sustainability transition. In this chapter we start with a brief overview of the history of food marketing before we talk about the changes that are needed from a social perspective as well as from the point of view of the individual consumer. We then elaborate what the new food marketing should be delivering to the consumer, namely support of ethical and sustainable products with transparency about the food production process and aimed at achieving population health and wellness. Food marketing needs to enter a new stage that responds to the planetary challenges and influences behaviour by making us look up and out, and most importantly beyond what is on our plate.

History of Food Marketing

In ancient times, finding food was a primal human preoccupation. This has changed enormously throughout the centuries with plant cultivation, raising domestic animals, developing subsistence and modern agriculture. Humans found ways to secure food which were easier, less laborious, less stressful and more efficient. The development of farming allowed for the production of surplus food which was exchanged through barter transactions and later on the market. Food became more than just a means for survival, it was bringing economic benefits and earning profit to those who could achieve high yields.

The advancement of industrialisation, technological innovation and mechanisation allowed farming to produce high agricultural outputs to feed the population but also to support a wide range of other industries, such as textiles, furniture making, construction of boats, manufacturing of musical instruments, cosmetics and most recently biofuels. From being a humble supplier of basic calories and nutrition, agriculture began to fuel an entire new industry, namely the food industry – a complex network of business activities that process crops and animal-based components to produce and deliver foodstuffs to the consumer. Many of these foodstuffs are developed in research laboratories in order to appeal to the taste of the consumer. They contain ingredients, such as sugar, salt and trans fats, that consumers find attractive and gradually start to prefer over natural products like fruits, vegetables, grains, legumes or nuts. The food industry in cooperation with supermarket chains has also found ways to artificially ripen fruits and vegetables or deliver them from across the globe to the consumer table all year-round. All these developments hardly take into account the health impacts and environmental footprint of foods. Instead, food marketing emerged as a way to convince consumers to eat the new products in order to support the financial viability of the food industry.

Food marketing takes the form of speech-based communication and non-speech-related activities (Pomeranz & Adler, 2015) which can occur at any stage within the food system between the farm gate and the consumer. Advertising is commonly used as a speech-based communication and these days it also extends overtly and covertly to social media platforms. The non-speech-related marketing is reflected in the price of the products, location in the store and position on the food shelves, information provision about content, place of origin and labelling. Food marketing evolves within the socio-economic, cultural, legal, political and technological environments of the countries and reflects the institutions and processes that are put in place (Kaynak, 2000). It reflects the level of economic development and the state of the market economy.

Back in the late 1800s, food marketing strategies were not common, as people were exchanging or purchasing, distributing and selling their food products locally due to convenience, lack of transportation infrastructure and the higher cost of transporting goods at that time (Son, 2015). Another example is the former Socialist Block where food marketing did not exist because the government-owned industries were not geared towards making profits. With economic development, food provision shifts from production to sales and then takes a consumer focus where issues related to quality regulations and traceability become important (Kaynak, 2000). The latter results in segmentation of the market (Murata, 2007) with tailored messages delivered initially directly, in the print media, radio and television and later on the Internet targeting the psychological profiles of different demographic, gender and market segments.

Many of the food companies operate beyond national borders and develop global strategies to attract loyal patrons and followers. National advertising campaigns are also employed to increase the visibility of particular products beyond those of their competitors who in some cases may be small local growers. Research evidence indicates that food advertising affects eating behaviour (Folkvord et al., 2016a). Individual predisposition or inclination factors determine different levels of susceptibility to food cues in advertisements (Valkenburg & Peter, 2013), with some population groups being more susceptible than others. For example, stronger effects are observed for individuals with a higher weight status (Norman et al., 2018), more impulsive nature (Hershko et al., 2019; Folkvord et al., 2016b), restrained eaters (Alblas et al., 2020) and children who are prohibited to consume certain foods, such as sweets and candies (Binder et al., 2020).

Food marketing is particularly effective in stimulating and reinforcing the consumption of particular categories of food, especially energy-dense products (Folkvord & Hermans, 2020). In this day and age, instead of supporting the public good, food marketing is used mainly to stimulate purchasing behaviour with little regard for consequences related to somebody's waistline or the natural environment. Finding solutions to the environment and climate emergency also means a shift in the marketing of food.

New Approaches in Food Marketing

Food marketing is occupying a space in the public realm which economists describe as non-excludable and non-avoidable. All consumers are subjected to speech-based and non-speech- related marketing with a very few options to avoid it. The assumption is that if a food company pays appropriately, it gets the right to expose people to its messages within a regulatory framework that does not reflect the challenges of the planetary emergency nor the priorities of human health. There are however many examples of government stepping in to protect human health (e.g. anti-smoking campaigns or COVID-19 vaccination) or natural resources (e.g. campaigns directed to reducing water use or improve waste collection).

It is no longer appropriate for food marketing to simply buy public space and encourage behaviours which are detrimental to the planet. Food marketing should not aim to connect with the consumers to enhance product/brand visibility and exploit human nature to achieve more sales (Kendall, 2014). It needs to re-position its role within the economy. Such changes are needed at two levels – social and individual. Below we discuss social marketing and personalised nutrition as two possible avenues for change together with some new methods to influence consumers.

Social Marketing

Social marketing refers to marketing that aims at behaviour changes that benefit larger sections of society and contribute to the greater common good (Dann, 2010; Gordon, 2011). An example of this is the "Go for 2&5" campaign in Australia promoting the consumption of two fruits and five vegetables per day (Carter et al., 2011). Social marketing can be positively or negatively framed, respectively encouraging or discouraging certain behaviours, or can aim at informing people by providing scientifically supported facts. For example, "Every Cigarette is Doing You Damage" campaigns run by the Departments of Health in Australia and New Zealand highlight the effects of smoking based on scientific evidence about the impact on different human organs, such as the aorta, emphysema or eyes, and communicate the Quitline telephone number. This example combines all three aspects by providing accurate information about the negative consequences of smoking while at the same time encouraging a proactive response to contact the Quitline to give up this addictive habit. Social marketing should become the norm and expectation not only on behalf of the government but also the private sector.

Sustainability social marketing (Bogueva et al., 2017) of food is needed in the new world of planetary emergency and pandemic conditions. It should help facilitate a transition to better food choices with the right nutrition in connection with care for the environment and an ethical side related to animal welfare. Such marketing campaigns run by the government are likely to have similar success as those related to smoking. Food companies, on the other hand, should endeavour to create special messages to build relationships for healthy eating and establish communities of like-minded individuals (Mintel, 2021) encouraging sustainability transitions.

Among the non-speech-related social marketing techniques are price alteration or discount strategies to provoke nudging (through a combination of "priming" and "salience") towards better food choices (Wilson et al., 2016). Labelling and information about what is inside the products, all its ingredients and their sourcing are becoming increasingly important as social marketing techniques. For example, the voluntary Health Star Rating in Australia and New Zealand assigning from half to five stars to packaged foods and beverages based on their nutritional value, is found to influence consumer choices to a certain degree (Talati et al., 2017). Food neophobia can also be eliminated with clear labelling of products (Alcorta et al., 2021). In line with the new COVID-19 reality, many people have started to make more serious commitments to reading food labels, be less lured toward oversized or multi-pack offers to reduce health risks associated with overeating and unhealthy eating.

Positioning products with better environmental and health credentials at eye-level on the shelves of grocery stores (Kendall, 2014) is another way to influence a broad range of consumers but it may need to be combined with in-store promotions to encourage purchases (Young et al., 2020). Social marketing should also keep up with the emerging food-purchasing behaviour of the younger generational cohorts, including the Millennials (Generation Y) and Centennials (Generation Z). Capitalising further on the COVID-19-imposed strengthening of the role of community, the marketing of food, beverages and food services should position them as caring for the common interests.

A specific form of social marketing is the production and dissemination of documentary films which expose facts about food production and consumption. The list is very long and in some cases these films are crowd-funded (e.g. *Cowspiracy*), in others they are sponsored by the new alternative proteins industry (e.g. *The Game Changers*), many are made by renown environmentalists (e.g. *Breaking Boundaries*) or sponsored by environmental entrepreneurs (e.g. *Seaspiracy*). Some commentators describe the power of such documentaries, including those revealing the conditions in which animals are being raised and slaughtered for food (e.g. *Earthlings*, *Unity* and *Food Inc.*) and impacts of food on human health (e.g. *What the Health* and *Forks over Knives*) as totally changing the lives of meat-eaters (Fletcher, 2020).

National dietary guidelines also play a role in social marketing and many limit the intake of red meat (e.g. Australia's 2013 dietary guidelines) because of its link to cancer. An analysis of 85 national food-based dietary guidelines however shows that the majority of them, namely 98%, were incompatible with at least one of the global health and environmental targets, including reduction in deaths related to chronic diseases, greenhouse gas emissions, freshwater use, cropland use and fertiliser application. It concludes that: "Providing clearer advice on limiting in most contexts the consumption of animal source foods, in particular beef and dairy, was found to have the greatest potential for increasing the environmental sustainability of dietary guidelines, whereas increasing the intake of whole grains, fruits and vegetables, nuts and seeds, and legumes, reducing the intake of red and processed meat, and highlighting the importance of attaining balanced energy intake and weight levels were associated with most of the additional health benefits" (Springmann et al., 2020, p. 1). Hence, social marketing is not properly reflected even at that high level of dietary recommendations leaving the door open for food companies to exploit the public space for individual profits at the expense of the common good.

Personalised Nutrition

At the other end of the marketing spectrum is personalised nutrition, also described as precision nutrition, tailored nutrition, stratified nutrition, nutrigenomics, nutrigenetics or nutritional genomics, which aims at improving human health by offering tailored solutions. Addressing increasingly changing lifestyles, genetic factors preventing the intake of certain foods (e.g. gluten-based, lactose-based and other allergens) and responding to gut flora conditions, personalised nutritional wants to deliver optimal health and wellness (van Delden et al., 2019). Technological and scientific advances, including in big data, allow a shift away from one-size-fits-all meals by offering customised diet plans. The expansion of lifestyle-imposed diseases like type 2 diabetes, heart disease and obesity, and the raising awareness about health benefits from dietary modifications are propelling the market growth of personalised nutrition – an entire new industry, forecast to reach a value of over $70 billion globally by 2025 (Swaney, n.d.). It combines insights from epigenomics, metabolomics and microbiomics to "preserve or increase health using genetic, phenotypic, medical, nutritional and other relevant information" (Ordovas et al., 2018, p. 1). Food marketing has the task to present a customised approach to anyone's genetic makeup with human DNA-tailored diets. This can be applied for dealing with particular health conditions or dietary requirements (e.g. for children or old-age groups) or for improving the overall population health.

The varieties of business segments companies are entering is so diverse, including specialised apps, offering personalised diet plans, supplements, gene testing and expert advice. Some examples of major players operating in the personalised nutrition market comprise the multi-level marketing company *Amway*, gut microbiomes testing businesses *Viome* and *Nutrigenomics*, customised vitamin packs company *Persona Nutrition* and personalised pharmaceuticals *Bayer AG* and *Bactolac Pharmaceutical Inc.*, and also the biotechnologies-oriented *Lonza*, DNA-based and evidence-backed supplement producers *Caligenix* and *GX Sciences*. Others like *Zipongo* are offering personalised meal plans, *Panaceutics Inc.* is producing customised nutritional products, while *Habit Food Personalized Inc.* is providing nutrition diet plan to the user's doorstep. *Mindbodygreen* is dedicated to helping consumers live their best life mentally, physically, spiritually, emotionally and environmentally. *Nutritional Genomix* is specialist advice-driven, offering experts in geneticists, bioinformaticians, dietitians, personal trainers, computer engineers and genetic counsellors, while *InsideTracker* is giving home kits for checking wellness. Actively participating in the manufacturing of fully personalised products is also the food industry which aims to capitalise on the opportunity by making strategic collaborations.

Personalised nutrition builds on individual genetic makeup to achieve optimal health and disease prevention and by doing so optimises the use of resources. Individual genetic testing provides information how the body processes and stores food and different nutritional components (Swaney, n.d.). However, some are sceptical about the full potential of personalised nutrition because of many more factors at play than genetic makeup (Swaney, n.d.) and express concerns about overpromising and creating a biohype (Stenne et al., 2012). Others believe that personalised nutrition could act as a disruptor to the current mass-production food systems (Askew, 2020) and eliminate food waste. However, it needs to be deployed in a way that does not increase the health inequalities within society and should build on the biological as well as lifestyle characteristics of the individuals (Ordovas et al., 2018).

One way is the development of targeted foods, meals and diets that deliver the precise health benefits the individual needs (van Delden et al., 2019). Another approach is the production of the core products at a large manufacturing scale and then adapting them to meet individual needs through assembling meal kits at a later stage (Askew, 2020) or making fresh meals with the help of specialised digital software (Vita Mojo, n.d.). A new route is collecting microbiome research-based data to inform the development of targeted nutritional advice, products or services to assist people achieve a lasting dietary change in behaviour that is beneficial for their health (Ordovas et al., 2018). Food marketing can play a role in offering special sensorial experiences and delivering nutritional interventions as solution suited to each individual and covering their beliefs, needs, body composition and ethics (Ordovas et al., 2018).

Preferences for functional foods, which offer benefits beyond nutrition and support general health by protecting against diseases and promoting healthy growth and development, are also a raising trend influenced by the search for personalised nutrition. A lot of these qualities are present in fruits and vegetables, but consumers also look for fortified foods which have increased content of vitamins, minerals, fibre, probiotics and antioxidants. They are expected to increase the level of energy, support better sleep, reduce stress, enhance focus, improve mood, help digestion, strengthen the immune system and facilitate specific functions of the human body (Kaur & Das, 2011; Kasemsap, 2018). The bigger interest in functional foods is from the Millennials and Generation Z (Centennials) who are extremely focused on diets backing both mental and physical wellbeing (Danley, 2021; IFT, 2021). An additional factor is the COVID-19 pandemic which boosts further interest in personalised nutrition and functional foods, particularly with plant-based ingredients. Consumers are open to unique flavours and experiences, but they are not ready to compromise on naturalness which creates a challenge for science and the food industry, with marketing needing to find a way to position personalised nutrition on the verge between natural and technology-driven way of satisfying dietary needs.

New Marketing Techniques

In the last decade, there have been a number of innovative marketing techniques used for food promotion. They are facilitated by the new social media platforms and the increased interest in health and wellbeing. Consumers are switching to mindful (consciously aware) and intuitive (listening to your body) eating that promote a healthy attitude towards food. They are increasingly using health apps to seek proof and incentives through the use of technology (Mintel, 2021). The interest in delivery apps that provide information about the food production story behind the scenes is also exponentially growing.

Direct marketing of food via the explosively growing social media, including Instagram, Twitter, Facebook, YouTube and TikTok, is becoming also popular due to its flexibility and easy engagement with sale offers. Clever social media campaigns are growing attracting millions of followers. For example, as of August 2021, Ben & Jerry's ice cream has 1.6 million followers on Instagram, 525,500 on Twitter and millions of views on YouTube. The American ice cream manufacturer also links its operations with corporate responsibility and values, such as climate change and renewable energy, racial justice and fair trade. Although the majority of its ice creams are dairy-based, 19 different vegan non-dairy options are offered and have the endorsement of the People for the Ethical Treatment of Animals (PETA).

Marketing also discovered the power of Autonomous Sensory Meridian Response (ASMR). Using sound and visuals of people biting, whispering, tapping, chewing and slurping different types of foods, ASMR videos and podcasts aim to generate a tingling sensation throughout the body, provoke a positive reaction and result in a purchase. This novel technique is adopted for goods associated with pleasure, such as chocolate or drinks, and it opens up a new era of communication and connection with the consumers filled with relaxed, positive messages. Using positive messages about the sustainability of our food choices can also influence people's attitudes and ways of thinking (Bogueva & Marinova, 2020).

Other forms of marketing are the use of characters, celebrities, cartoons, animations and brand iconography that create a lasting impression and develop a relationship with the consumers. In their effort to attract especially children, food marketing has used toys, collectible cards, games, competitions, children's clubs, gifts and other prizes. Such strategies have been criticised for adding to the plastics pollution problems as has been the case with the Australian Coles supermarket (Cox, 2021). This supermarket consequently transitioned to more sustainability practices, such as the use of 100% renewable energy, reducing plastic packaging and elimination of food waste (Coles, 2021). An interesting alternative approach is Woolworths' Fresh

Food Kids' Discovery Garden (Woolworths, 2021) with real vegetables, flowers and herbs that children can plant and meaningfully engage with their natural environment. Food marketing is thus responding to changing consumer expectations and we look at this in further detail below.

Expectations from Food Marketing

The consumers these days are not passive acceptors of the messages from food marketing. They want to engage with the matter and anticipate particular values and certain behaviour in the marketing of food. This is how they want to regain control of the public space by contributing to setting the rules of the marketing game. Marketing needs to be ethical in the subject matter it engages with, should take account of issues related to the natural environment and contribute towards health and wellbeing.

Ethical Marketing

Ethical expectations among consumers are on the rise. The ethics of food consumption pertains to the moral consequences of consumer choices in relation to the inhumane treatment of animals and animal welfare but also to other equally important concerns, such as environmental damage, the use of industrialised farming methods with chemical pesticides and artificial fertilisers affecting biodiversity and soils, the unfairness in exploitive labour practices and unfair pay, the unequal distribution of food causing shortages around the world (Food Ethics Council, 2021). As a growing phenomenon moving faster into the mainstream (Deloite, 2019; Food Frontier, 2021), plant-based food choices and alternatives to animal source proteins (see Fig. 8.1) are also related to the ethical expectations of the consumers.

In addition to food description as "fresh", "artisan" and "authentic", consumers are turning towards more specific and appealing terms which describe how the food has been grown, such as "organic", "non-GMO", "home-grown", "non-dairy", "plant-based" and "free-range" (Deloite, 2016; Olayanju, 2019; Lawrence & King, 2019). Consumers are increasingly interested in detailed labelling for transparency and traceability of ingredients where the application of blockchain technology has already commenced. The disruptions caused by COVID-19 ignited interest in the supply chain and provenance of food and its components (Nielsen, 2021). Brands have the opportunity to gain market share by providing increased transparency about their products (Label Insight, 2016) and by using ethically sourced ingredients. Digital technology is further facilitating this trend by delivering instant information to the interested consumers.

Fig. 8.1 Alternative proteins and packaging. (Source: Authors)

Ethical food marketing should also avoid stereotyping people and their food choices as has been the case with meat considered as a symbol of masculinity (Bogueva & Marinova, 2019). A transformative change is needed in a climate emergency where old narratives are no longer appropriate and the attention should explicitly focus on creating positive advertising messages of environmentally sustainable and ethical food.

Environmentally Responsible Marketing

Like the production of cigarettes or the building of weapons, the creation of food is a very large and lucrative business. The welfare of the natural environment is often forgotten. Soil health, water use, greenhouse gas emissions, plastics pollution, antimicrobial resistance and biodiversity loss seem to exist in a parallel universe to food production. However, consumers are increasingly becoming aware of the challenges threatening our planet and are demanding food that is nutritious and produced sustainably. They expect food marketing to be environmentally responsible and not to promote products that are not sustainable.

Some companies started to build their image around sustainability features of their products, such as Corona becoming the first global beverage brand with a net zero plastic footprint (Kazarian, 2021) or PepsiCo, Kraft Heinz and Nestlé developing water strategies and awareness project initiations (Shilling, 2020). Sustainable food is considered safe, produced without hazardous pesticides and chemicals damaging the land and water, and healthy without non-essential antibiotics or added growth hormones. A consumer trend is the desired food to be produced in an environmentally friendly and sustainable way and by producers with a commitment to the natural environment, using farming technologies that seek to reduce the impact on natural resources (IFIC, 2020, p. 10). This is the type of food that marketing should endorse. This will be of benefit to the natural environment whilst satisfying consumer demand.

Marketing for Health and Wellness

The food marketing methods have come under scrutiny because of encouraging poor dietary choices. They are not only affecting the well-being of people but also public health budgets. Lately there has been a turn around with marketing helping consumers to make healthy choices.

Consumers are progressively turning into food that is treating health conditions, providing multiple benefits through vitamins, minerals and prebiotics and supporting overall immunity (Sloan, 2021). The majority of these foods are plant-based and have the added benefits of low environmental impact. Another similar trend is related to fermented foods and drinks produced by yeast and bacterial fermentation, including yogurts, kefir, kimchi, sauerkraut, pickled vegetables, vinegar, kombucha and sourdough, which enhance the nutritional quality of food (Poutanen et al., 2009; Borresen et al., 2012; Kok & Hutkins, 2018). The consumption of fermented foods is considered as a panacea, linked with many health advantages, including minimised risks of type 2 diabetes, heart disease and weight management (Kok & Hutkins, 2018). Such health benefits are achieved because of the probiotic properties of naturally available microorganisms, such as the yogurt starter culture organisms Lactobacillus delbrueckii subsp. bulgaricus, Streptococcus thermophilus, Bifidobacterium and Lactobacillus strains (Bogueva et al., 2021).

A new trend that has grown among consumers is associated with food products that positively affect the mood, physical, mental and emotional state of consumers. Again, the majority of these products are plant-based and they help against insomnia, tension, reducing stress, boosting the brain, improving mental health and skincare. Examples include beverages like Recess which claim to have mood-healing properties, such as feeling calm despite the stressful world around you and creating a moment to think about things that matter, or the Australian brand Shine+ beverages containing nootropics, ingredients that improve the brain function, including L-theanine, caffeine, gingko biloba, turmeric, B vitamins and green tea.

The use of various botanical representatives, such as echinacea, elderberry, turmeric, cranberry, ivy leaf, ginger, garlic and green tea, is also a leading trend. Novel alternative formulations that help consumers to avoid unwanted ingredients and perceived allergens are also on the consumer radar (Sloan, 2021). Some are searching for food that corresponds to their lifestyle diet, such as vegan, vegetarian or flexitarian. Such choices can be supported by a health monitoring device or app, related to the specific dietary guidance, food consumption, exercise plans, physical activity or overall health (IFIC, 2020). All these trends also align with personal nutritional, physical health improvement or ethical goals.

Many novel food technologies evoke negative reactions among consumers and their societal acceptance is often shaped by risk perception, unclear benefits, lack of trust in the food industry, citizen knowledge, attitudes and cost (Gupta et al., 2012; McClements, 2019) as well as by consumer personality characteristics, such as technology disbelief, food neophobia, food disgust and sensitivity (Ruzgys & Pickering, 2020). Marketing efforts should be directed toward finding alternative pathways for human development – physical, intellectual, emotional and spiritual, in relation to nutrition from economic, social, environmental and biological perspectives to offer mental and emotional wellbeing solutions creating a new foundation for healthy eating (Mintel, 2021).

We urgently need to be shifting from food marketing to nutrition marketing. Increasingly, consumers are becoming aware and willing to look at the right nutrition and redirecting their focus from previously advertised low-fat and reduced calories food to healthy and nutritious functional and personalised food. Nowadays, and especially in the future, consumers would like their food to be created for their lifestyle needs with relevance to their ethics, activities in daily life and the natural environment.

Conclusion

If we try to understand the ubiquitour role of food marketing, we need to turn back to history. It all commenced with the progressive human mind and actions and their ability to secure food. The resultant overproduction required finding ways to utilise the overabundance of calories introducing it into the food industry which encouraged people to replace the plant-based options with processed products and eat more. This was a way for the food industry to secure profits and market influence (Nestle, 2013).

Similar to the "Look up" advertising campaign, food marketing needs to also look up and out and see the real world. It needs to engage with the reality of the planetary emergency and look beyond the computer screens with profit-and-loss statements. Over the years, food marketing has contributed to promoting products with high environmental and health price tags. It has helped spread the Western type diet, high on meat and processed foods, all over the globe. The responsibility has been to the one who pays for its services and not to the consumers or the planet. Governments and civil society have not been able to stop this operational model, in fact, they have not even attempted to protect the public space.

What the food industry and its marketing have created over the years must be completely reversed today. Now, food marketing is tasked with the job to make all of us look up. This is not a difficult task given the changes that are already occurring in society with greater awareness about health and environmental problems and with higher expectations. The change in people's eating behaviour is increasingly urgent and whilst many people are already opening their eyes and minds to foods with fewer fats, carbohydrates and less processing, food marketing needs to embrace and encourage such transitions. Food marketing has to enter a new, next stage, one that creates social goods and does not destroy the planet. As a unique facet of human connection (Kerr & Maze, 2019), hope will excite consumers and this is what they will want and demand.

References

Adams, C. J. (1990). *The sexual politics of meat: A feminist-vegetarian critical theory*. Bloomsbury.

Alblas, M. C., Mollen, S., Fransen, M. L., & van den Putte, B. (2020). Food at first sight: Visual attention to palatable food cues on TV and subsequent unhealthy food intake in unsuccessful restrained eaters. *Appetite, 147*, 104574. https://doi.org/10.1016/j.appet.2019.104574

Alcorta, A., Porta, A., Tárrega, A., Alvarez, M. D., & Vaquero, M. P. (2021). Foods for plant-based diets: Challenges and innovations. *Food, 10*, 293. https://doi.org/10.3390/foods10020293

Askew, K. (2020). *Personalised nutrition will shape the food systems of the future: How will mass production adapt*. Food Navigator. https://www.foodnavigator.com/Article/2020/06/24/Personalised-nutrition-will-shape-the-food-systems-of-the-future-How-will-mass-production-adapt#:~:text=Personalised%20nutrition%20is%20a%20phrase,deliver%20optimum%20health%20and%20wellness

Binder, A., Naderer, B., & Matthes, J. A. (2020). "Forbidden fruit effect": An eye-tracking study on children's visual attention to food marketing. *International Journal of Environmental Research and Public Health, 17*(6), 1859. https://doi.org/10.3390/ijerph17061859

Bogueva, D., & Marinova, D. (2019). Reconciling not eating meat and masculinity in the marketing discourse for new food alternatives. In D. Bogueva, D. Marinova, T. Raphaely, & K. Schmidinger (Eds.), *Environmental, health and business opportunities in the new meat alternatives market* (pp. 260–282). IGI Global.

Bogueva, D., & Marinova, D. (2020). Autonomous Sensory Meridian Response (ASMR) for responding to climate change. *Sustainability, 12*, 6947. https://doi.org/10.3390/su121769

Bogueva, D., Raphaely, T., Marinova, D., & Marinova, M. (2017). Sustainability social marketing. In J. Hartz-Karp, D. Marinova, & D. (Eds.), *Methods for sustainability research* (pp. 280–291). Edward Elgar.

Bogueva, D., Apostolova, M., & Danova, S. (2021). Food, nutrition and health in Bulgaria. In A. Gostin, D. Bogueva, & V. Kakurinov (Eds.), *Nutritional and health aspects of food in the Balkans* (pp. 67–89). Elsevier.

Borresen, E. C., Henderson, A. J., Kumar, A., Weir, T. L., & Ryan, E. P. (2012). Fermented foods: Patented approaches and formulations for nutritional supplementation and health promotion. *Recent Patents on Food, Nutrition & Agriculture, 4*(2), 134–140. https://doi.org/10.2174/2212798411204020134

Carter, O. B., Pollard, C., Atkins, J. F., Milliner, J. M., & Pratt, I. S. (2011). 'We're not told why – We're just told': Qualitative reflections about the Western Australian 'Go for 2&5' fruit and vegetable campaign. *Public Health Nutrition, 14*(6), 982–988. https://doi.org/10.1017/S1368980010003381

Coles. (2021). *Coles' bold new ambitions to reduce its environmental impact*. https://www.coles.com.au/whats-happening/sustainability/coles-bold-new-ambitions-to-reduce-its-environmental-impact

Cox, L. (2021). *Coles bins Stikeez and minis for good following criticism of plastic promotions*. The Guardian. https://www.theguardian.com/business/2021/jul/23/coles-bins-stikeez-and-minis-for-good-following-criticism-of-plastic-promotions

Danley, S. (2021). *Shifting demographics poise functional products for growth*. Food Business News. https://www.foodbusinessnews.net/articles/18343-shifting-demographics-poise-functional-products-for-growth

Dann, S. (2010). Redefining social marketing with contemporary commercial marketing definitions. *Journal of Business Research, 63*(2), 147–153. https://doi.org/10.1016/j.jbusres.2009.02.013

Deloitte. (2016). *Capitalising on the shifting consumer food value equation*. https://www2.deloitte.com/content/dam/Deloitte/us/Documents/consumer-business/us-fmi-gma-report.pdf

Deloitte. (2019). *Plant-based alternatives: Driving industry M&A*. https://www2.deloitte.com/content/dam/Deloitte/uk/Documents/consumer-business/deloitte-uk-plant-based-alternatives.pdf

Fletcher, H. (2020). *The 10 films sure to turn meat-eaters vegan*. Evening Standard. https://www.standard.co.uk/culture/film/the-10-films-sure-to-turn-meateaters-vegan-a3743351.html#comments-area

Folkvord, F., & Hermans, R. C. J. (2020). Food marketing in an obesogenic environment: A narrative overview of the potential of healthy food promotion to children and adults. *Current Addiction Reports, 7*, 431–436. https://doi.org/10.1007/s40429-020-00338-4

Folkvord, F., Anschütz, D. J., Boyland, E., Kelly, B., & Buijzen, M. (2016a). Food advertising and eating behavior in children. *Current Opinion in Behavioral Sciences, 9*, 26–31. https://doi.org/10.1016/j.cobeha.2015.11.016

Folkvord, F., Veling, H., & Hoeken, H. (2016b). Targeting implicit approach reactions to snack food in children: Effects on intake. *Health Psychology, 35*(8), 919–922. https://doi.org/10.1037/hea0000365

Food Ethics Council. (2021). *What is food ethics?* https://www.foodethicscouncil.org/learn/food-ethics/what-is-food-ethics/

Food Frontier. (2021). *2020 state of the industry*. https://www.foodfrontier.org/wp-content/uploads/dlm_uploads/2021/03/Food-Frontier-2020-State-of-the-Industry.pdf

Ganry, J., Egal, F., & Taylor, M. (2011). Fruits and vegetables: A neglected wealth in developing countries. *Acta Horticulturae, 921*, 105–109. https://doi.org/10.17660/ActaHortic.2011.921.12

Gordon, R. (2011). Critical social marketing: Definition, application and domain. *Journal of Social Marketing, 1*(2), 82–99. https://doi.org/10.1108/20426761111141850

Gupta, N., Fischer, A. R., & Frewer, L. J. (2012). Socio-psychological determinants of public acceptance of technologies: A review. *Public Understanding of Science, 21*(7), 782–795. https://doi.org/10.1177/0963662510392485

Hershko, S., Cortese, S., Ert, E., Aronis, A., Maeir, A., & Pollak, Y. (2019). Advertising influences food choices of university students with ADHD. *Journal of Attention Disorders, 25*(8), 1170–1176. https://doi.org/10.1177/1087054719886353

Institute of Food Technologists (IFT). (2021). *9 food trends to watch for in 2021*. https://www.ift.org/news-and-publications/blog/2021/2021-food-tech-predictions

International Food Information Council (IFIC). (2020). *2020 food & health survey*. https://foodinsight.org/wp-content/uploads/2020/06/IFIC-Food-and-Health-Survey-2020.pdf

Kasemsap, K. (2018). Nutrition and functional foods: Current trends and issues. In A. Verma, K. Srivastava, S. Singh, & H. Singh (Eds.), *Nutraceuticals and innovative food products for healthy living and preventive care* (pp. 158–175). IGI Global.

Kaur, S., & Das, M. (2011). Functional food: An overview. *Food Science and Biotechnology, 20*(4), 861–875. https://doi.org/10.1007/s10068-011-0121-7

Kaynak, E. (2000). Cross-national and cross-cultural issues in food marketing: Past, present and future. *Journal of International Food & Agribusiness Marketing, 10*(4), 1–11. https://doi.org/10.1300/J047v10n04_01

Kazarian, K. (2021). *Corona is the first global beverage brand to reach net zero plastic footprint*. https://www.foodengineeringmag.com/articles/99501-corona-is-first-global-beverage-brand-to-reach-net-zero-plastic-footprint

Kendall, G. (2014). *The science that makes us spend more in supermarkets, and feel good while we do it*. The Conversation. https://theconversation.com/the-science-that-makes-us-spend-more-in-supermarkets-and-feel-good-while-we-do-it-23857

Kerr, F., & Maze, L. (2019). *The art & science of looking up: Transforming our brains, bodies, relationships and experience of the world by the simple act of looking up.* http://www.gliderglobal.com/wp-content/uploads//THE-ART-AND-SCIENCE-OF-LOOKING-UP-REPORT_2019_SINGLES.pdf

Kok, C. R., & Hutkins, R. (2018). Yogurt and other fermented foods as sources of health-promoting bacteria. *Nutrition Reviews, 76*(s1), 4–15. https://doi.org/10.1093/nutrit/nuy056

Label Insight. (2016). *How consumer demand for transparency is shaping the food industry.* https://www.labelinsight.com/hubfs/Label_Insight-Food-Revolution-Study.pdf?hsCtaTracking=fc71fa82-7e0b-4b05-b2b4-de1ade992d33%7C95a8befc-d0cc-4b8b-8102-529d937eb427,%203

Lawrence, S., & King, T. (2019). *Meat the alternative: Australia's $3 billion dollar opportunity.* Food Frontier. https://www.foodfrontier.org/wp-content/uploads/2019/09/Meat-the-Alternative-Food-Frontier.pdf

Look Up. (n.d.). *What happened when Australia looked up?* https://www.lookup.org.au/the-effect

McClements, D. J. (2019). *Future food: How modern science is transforming the way we eat.* Springer Nature.

Mintel. (2021). *Global food and drink trends 2021.* https://www.mintel.com/global-food-and-drink-trends

Murata, Y. (2007). Taste heterogeneity and the scale of production: Fragmentation, unification, and segmentation. *Journal of Urban Economics, 62,* 135–160. https://doi.org/10.1016/j.jue.2006.11.005

Nestle, M. (2013). *Food politics: How the industry influences nutrition and health* (3rd ed.). University of California Press.

Nielsen, B. (2021). *Chain of custody: Overcoming food supply chain disruption.* Grande. https://www.grandecig.com/blog/food-supply-chain-of-custody

Norman, J., Kelly, B., McMahon, A. T., Boyland, E., Baur, L. A., Chapman, K., … Bauman, A. (2018). Sustained impact of energy-dense TV and online food advertising on children's dietary intake: A within-subject, randomised, crossover, counter-balanced trial. *International Journal of Behavioural Nutrition and Physical Activity, 15*(1), 37. https://doi.org/10.1186/s12966-018-0672-6

Olayanju, J. B. (2019). *Top trends driving change in the food industry.* Forbes. https://www.forbes.com/sites/juliabolayanju/2019/02/16/top-trends-driving-change-in-the-food-industry/?sh=1231e6c66063

Ordovas, J. M., Ferguson, L. R., Tai, E. S., & Mathers, J. S. (2018). Personalised nutrition and health. *British Medical Journal (BMJ), 361,* k2173. https://doi.org/10.1136/bmj.k2173

Pollan, M. (2006). *The omnivore's dilemma: The search for a perfect meal in a fast-food world.* Bloomsbury.

Pomeranz, J. L., & Adler, S. (2015). Defining commercial speech in the context of food marketing. *The Journal of Law, Medicine & Ethics, 43*(s1), 40–43. https://doi.org/10.1111/jlme.12213

Poutanen, K., Flander, L., & Katina, K. (2009). Sourdough and cereal fermentation in a nutritional perspective. *Food Microbiology, 26*(7), 693–699. https://doi.org/10.1016/j.fm.2009.07.011

Ruzgys, S., & Pickering, G. J. (2020). Perceptions of cultured meat among youth and messaging strategies. *Frontiers, Sustainable Food Systems, 4,* 122. https://doi.org/10.3389/fsufs.2020.00122

Schlosser, E. (2002). *Fast food nation: The dark side of the all-American meal.* Perennial.

Shilling, R. (2020). *How two industry leaders invest to save the water supply as scarcity grows.* Food Engineering Magazine. https://www.foodengineeringmag.com/articles/99163-how-two-industry-leaders-invest-to-save-the-water-supply-as-scarcity-grows

Sloan, E. (2021). *Top 10 food trends in 2021.* Technology Magazine, IFT. https://www.ift.org/news-and-publications/food-technology-magazine/issues/2021/april/features/top-10-food-trends-of-2021

Smith, B. (2020). *The 17 best display ads of 2020 (and why they work).* Word Stream. https://www.wordstream.com/blog/ws/2020/09/30/best-display-ad-examples

Son, H. (2015). The history of Western futures studies: An exploration of the intellectual traditions and three-phase periodization. *Futures, 66*, 120–137. https://doi.org/10.1016/j.futures.2014.12.013

Springmann, M., Spajic, L., Clark, M. A., Poore, J., Herforth, A., Webb, P., Rayner, M., & Scarborough, P. (2020). The healthiness and sustainability of national and global food based dietary guidelines: Modelling study. *British Medical Journal (BMJ), 370*, m2322. https://doi.org/10.1136/bmj.m2322

Stenne, R., Hurlimann, T., & Godard, B. (2012). Are research papers reporting results from nutrigenetics clinical research a potential source of biohype? *Accountability of Research, 19*(5), 285–307. https://doi.org/10.1080/08989621.2012.718681

Swaney, S. (n.d.). *Personalised nutrition: What is it and does it really work?* Nutrition Australia. https://nutritionaustralia.org/division/nsw/personalised-nutrition-what-is-it-and-does-it-really-work/

Talati, Z., Norman, R., Pettigrew, S., Neal, B., Kelly, B., Dixon, H., … Shilton, T. (2017). The impact of interpretive and reductive front-of-pack labels on food choice and willingness to pay. *International Journal of Behavioral Nutrition and Physical Activity, 14*, 171. https://doi.org/10.1186/s12966-017-0628-2

Valkenburg, P. M., & Peter, J. (2013). The differential susceptibility to media effects model. *Journal of Communication, 63*(2), 221–243. https://doi.org/10.1111/jcom.12024

Van Delden, B., Aley, G., & Hogarth-Scott, P. (2019). *Personalised nutrition offering vast opportunity*. KPMG. https://home.kpmg/au/en/home/insights/2019/02/evokeag-personalised-nutrition-offering-vast-opportunity.html

Vita Mojo. (n.d.). *Our products: Smarter digital ordering and back of house solutions*. https://www.vitamojo.com/our-products

Wilson, A. L., Buckley, E., Buckley, J. D., & Bogomolova, S. (2016). Nudging healthier food and beverage choices through salience and priming. Evidence from a systematic review. *Food Quality and Preference, 51*, 47–64. https://doi.org/10.1016/j.foodqual.2016.02.009

Woolworths. (2021). *Woolworths' Discovery Garden*. https://www.woolworths.com.au/shop/discover/garden

Young, L., Rosin, M., Jiang, Y., Grey, J., Vandevijvere, S., Waterlander, W., & Mhurchu, C. N. (2020). The effect of a shelf placement intervention on sales of healthier and less healthy breakfast cereals in supermarkets: A co-designed pilot study. *Social Science & Medicine, 266*, 113337. https://doi.org/10.1016/j.socscimed.2020.113337

Zaraska, M. (2016). *Meathooked: The history and science of our 2.5-million-year obsession with meat*. Basic Books.

Part III
Individual Perspectives

Chapter 9
Flexitarianism

Abstract Flexitarianism – a word combining "flexible" and "vegetarianism", has found its way into English since the early 2000s. After introducing flexitarianism, the chapter discusses sustainable diets and why human eating habits, particularly in the West, need to change. It offers ideas how to mainstream flexitarianism so that plant-based food choices become the norm, rather than the exception, in the pursuit for a healthy and nutritional human diet. Mainstreaming flexitarianism is a way to link human food with the broad spectrum of environmental and social problems which are at the core of the current planetary emergency.

The word "flexitarianism" – a combination between "flexible" and "vegetarianism", has found its way into the modern English language since the early 2000s. It describes a western diet which aims to reduce meat intake. In 2003, "flexitarian" was voted the most useful new word of the year by the American Dialect Society (2004) and a decade later it was also listed in the popular Merriam-Webster's Collegiate dictionary (Italie, 2012) and the Oxford English Dictionary (Derbyshire, 2017). If you have not heard it until now, it is most likely because you do not live in a western country, your diet is not influenced by the West or you have not been interested until now in the impact human dietary choices have on climate change and the natural environment.

Explained simply, flexitarians are people who are conscientiously reducing the meat intake in their diet. The two main reasons for them deciding to do so are concerns about the natural environment and the way animals are exploited and killed for human consumption. Therefore, flexitarians are also described as ethical vegetarians or semi-vegetarians (Derbyshire, 2017). Behind this neologism lies a complex set of issues which we try to disentangle in the chapter. The final message is that the majority of people prefer not to be labelled or singled out for following a diet that socially distinguishes them from their circle of friends, peers, colleagues and relatives. Nevertheless, we are witnessing a gradual adoption of flexitarian habits among many sections of society in the West in order to respond to the climate emergency. In China – the world's largest consumer of red meat in absolute terms,

flexitarianism is not used as a description of a diet, but there are deliberate government efforts put in place to reduce the consumption of animal proteins to protect the
health of the country's population and the global environment.

The chapter starts with defining flexitarianism, then moves to a discussion about
sustainable diets and why human eating habits need to change. We conclude with
ideas how to mainstream flexitarianism so that plant-based food choices become the
norm rather than the exception for a healthy and nutritional human diet in the current planetary emergency. It is important to stress that there is a myriad of diets and
nutritional specialists who advocate particular approaches to eating; what we are
trying to achieve is link human food with the broad spectrum of environmental and
social problems which are at the core of the planetary emergency.

Defining Flexitarians

We are both originally from Bulgaria and under the traditional dietary guidance by
the Bulgarian orthodox religion – adopted in the country more than 11 centuries ago
(in 870), out of one calendar year 200 days people are supposed to abstain from
consuming animal-based foods, including meat, dairy and eggs (Vitarama, 2020;
Bogueva et al., 2021a, c). Such long periods of this type of fasting or lent are associated with religious celebrations but they also reflect the availability and seasonality
of food as well as the relationships within the community and the links between
people and nature. For example, there is a 40-day lent before Christmas when there
is a lot of different varieties of vegetables and grains available to be consumed. The
Great Lent before Easter is even longer – 49 days, and this is the time when people
finish eating off nuts and dried fruits as well as grains, lentils, beans and other
legumes from the previous harvest. The time of abstaining from animal-based foods
is also seen understood as a cleansing period which allows the human body to purify
itself, rejuvenate and spiritual values of kindness and compassion to be strengthened.

In Bulgaria we were raised as semi-vegetarians. This however was not because
of religious convictions, but due to restricted availability of animal-based food during socialist times. Also, the price of meat, milk, cheese, eggs and other animal-
based products in Bulgaria at that time reflected much better the high costs associated
with their production, including environmental restrictions linked to land availability and the labour intensity of livestock in the relatively small cooperatives in the
Bulgarian countryside and villages. A standard Bulgarian meal at that time would
include only a very small portion of meat or mince, around 50 g, with a lot of vegetables, legumes, rice and was always consumed with mainly wholemeal bread. It
was a normal practice meat to be eaten only once per day and not on a daily basis.
Industrial scale poultry production was introduced in the 1970s and was largely
resented by the population because of the inferior taste of the chicken meat and ethical concerns about the living conditions of the birds.

These days, under western influences, globalised food supply chains and technological advancements in processing, preservation and transportation of food, even in

Bulgaria people are no longer following healthy dietary patterns with low consumption of animal-based foods. Although meat consumption continues to be relatively low at 58 kg per person per year (as in 2017, Ritchie & Roser, 2019) compared to 124 kg per person per year in USA, Bulgarians have a strong preference for feta cheese, yogurt and other dairy products (Kostova et al., 2018). Flexitarianism is still a largely unfamiliar concept in the country and this small 7-million nation and its diaspora around the world are part of the big global nutritional shift that is driving climate change, overexploitation and pollution of land and natural resources as well as impacting on human health.

What is then flexitarianism? The flexitarian diet was first described by the dietitian Dawn Blatner (2009) as combining the benefits of "vegetarian eating while still enjoying animal products in moderation" (Streit, 2019, para. 5). Flexitarians do not abstain from meat, dairy, eggs, honey and other animal products; instead they emphasise the consumption of healthy plant-based foods as the main source of protein, vitamins and other nutrients. In many aspects, this way of eating is similar to the Mediterranean diet which has been seen as the healthiest choice for many years (US News Staff, 2021). It is assumed that a flexitarian diet is overall healthier than a standard meat-rich western diet and also has a lighter footprint on the natural environment. These claims are difficult to quantify or measure because of the actual flexibility of the diet which does not provide any particular recommendations or restrictions in terms of different food groups. Nevertheless, a US panel of recognised experts in nutrition, obesity, heart disease, diabetes and food psychology consistently rank the flexitarian diet high overall – shared first place with Mediterranean diet in 2020 (Doheny, 2020) and second in 2021 (US News Staff, 2021). This diet is also considered as easy to follow for the entire family (US News Staff, 2021).

Despite some distinguishing between beginners (6–8 meatless out of 21 meals per week), advanced (9–14 meatless out of 21 meals per week) and expert (more than 15 meatless out of 21 meals per week) flexitarians (Taub-Dix, 2019), a flexitarian diet does not aim to convert people into vegetarians or vegans. According to Raphaely et al. (2013), the aim of a flexitarian diet should be to follow the dietary guidelines produced by reputable organisations whose main concern is human health and to not exceed the restrictions imposed specifically on red meat. For example, the World Cancer Research Fund and the American Institute for Cancer Research in their Continuous Update Project state: "If you eat red meat [such as beef, veal, pork, lamb, goat and non-bird], limit consumption to no more than about three portions per week. Three portions are equivalent to about 350–500 g (about 12–18oz) cooked weight. Consume very little, if any, processed meat" (WCR & AICR, 2018, p. 29). This recommendation is endorsed in the UK Government while the Australian Dietary Guidelines limit red meat intake to 455 g of lean red meat per week (one serve of 65 g per day) with processed meats included in the nutrient poor discretionary foods which should be avoided (NHMRC, 2013). In 2019, the Australian Heart Foundation revised down its guidelines for red meat intake by limiting it to 350 g per week (Sparkes, 2019). The European Union's Guidelines consistently advise to eat red meat sparingly (European Commission, 2020). China also advises its citizens to limit their red meat consumption to 40–75 g per day, half of its current level, to avoid an impending public health crisis driven by shifting dietary preferences and changing lifestyles (Froggatt & Wellesley, 2016). At a global level, projected estimates show that adopting a flexitarian diet (which includes a maximum of 43 g/day of poultry and lamb, but not red meat) would result in 2030 in avoiding 12.7 million deaths related to four groups of non-communicable diseases, namely coronary heart disease, stroke, cancer and type 2 diabetes mellitus (FAO, IFAD, UNICEF, WFP & WHO, 2020). The majority of preventable deaths (74% or 9.4 million) are in middle-income countries (FAO, IFAD, UNICEF, WFP & WHO, 2020). These are also the countries with increasing westernisation of their diets where flexitarianism can counter this trend.

All of the above dietary limits are based entirely on health grounds. Although human health is an important consideration, the planetary emergency imposes other priorities that need to be taken into account when examining people's diets, particularly in places where there is ample food availability. In 2018, the EAT-Lancet Commission (Willett et al., 2019) was the first to combine human and ecological benefits and set global targets for 2015. Its recommendations respond to the planetary emergency and the role food plays in biodiversity loss, land use changes, water use, pollution of air, water and soils. Changing the ways we produce and consume food can also be the solution. The EAT-Lancet Commission put an overall target for sustainable food production that it should not use any additional land than the one currently dedicated to agriculture in order to safeguard biodiversity, decrease water consumption, reduce nitrogen and phosphorus pollution, achieve zero CO_2 emissions, no further increase in methane and nitrous oxide. It also recommended an overall 50% reduction in red meat consumption and doubling the amounts of fruits, vegetables, nuts and legumes eaten (Willett et al., 2019).

Although the EAT-Lancet Commission did not use flexitarianism anywhere in its report, what it argued about was fully in line with a flexitarian diet. It is estimated that as in 2017 such a dietary transition is relatively simply to achieve for 6 billion people on this planet while for 1.6 billion it would be unaffordable (Hirvonen et al., 2020). Such findings do raise issues of equity but a way to look at this is by revising the prices, subsidies and taxes on food rather than bluntly reject the recommendations. Our Bulgarian experience shows that higher prices for animal-based foods do influence people's preferences towards healthier dietary choices and also reflect better the planetary costs of human nutrition. Furthermore, the rich countries have more power in competing for grains on the global markets which are then inefficiently used as livestock feed rather than directly for human consumption. If 6 billion people can decrease their meat consumption this would inevitably lower down the prices of many plant-based foods making them much more affordable for poorer people.

A major aspect of flexitarianism is that it is a voluntary conscious decision. Concerns about the natural environment and animal welfare play a major role in this as do co-benefits related to human health, such as reduced diabetes risk, lower blood pressure, weight loss, prevention and treatment of inflammatory bowel disease (Derbyshire, 2017). A range of other factors also influence people's decisions to reduce their intake of animal-based foods. They include the Veganuary and Meatfree

Mondays campaigns, but also the COVID-19 pandemic with the disruption of meat supply chains and higher rates of infection in slaughter houses (Grant & Richter, 2020). The market supported by food research and development is also providing new animal-free products which have become available in supermarkets and restaurants. In 2018, it was estimated that the share of flexitarians was 34% in UK and two-thirds of people in USA were cutting their meat consumption (Grant & Richter, 2020). We personally adhere to a diet described as vegetarian which sits between veganism and vegetarianism with a very low and sporadic intake of products, such as yogurt, cheese, eggs and honey. The raising popularity of veganism (Starostinetskaya, 2020) will similarly contribute towards decreasing global consumption of animal-based foods. Irrespective as to how you describe your diet, consumers' ethical values, including sustainability, are expected to drive the food sector in the new COVID-normal and beyond (Mintel, 2021).

How we define or describe flexitarianism in a planetary emergency is much less important than the fact that people are conscientiously changing their diets towards decreasing the consumption of animal-based foods. Whether these changes will be substantial and fast enough to preserve the natural resources of this planet and to protect populations from the changing climate and weather patterns is a different issue which requires a much higher attention to the sustainability of human diets.

Sustainable Diets

A planetary health approach as the one espoused by the EAT-Lancet Commission is what is needed to address the environment and climate emergency our planet and its current and future human populations face. Human health can no longer be isolated from the health of ecosystems (Burlingame & Dernini, 2012). Restricting thinking around diets only to improve and maintain human health has been detrimental to the well-being of other life on this planet. There is overwhelming evidence that a western style of meat-rich diet is the worst choice in a planetary emergency. For example, a meta-analysis of 570 studies from 119 countries examining the life-cycle environmental footprints of 40 different food items which provide 90% of the global protein and calorie intake, shows that beef is the most environmentally detrimental option (Poore and Nemecek, 2018). Other studies which compare meat with plant-based options include Clark and Tilman's (2017) meta-analysis of 742 different production systems concluding that the emissions form ruminant meats are 20–100 times higher, and the analysis by Eshel et al. (2014) of the burden of the US diet which shows that per unit of calorie, beef requires respectively 4, 8 and 40 times more irrigated water than rice, potato and wheat. This supports the need for flexitarianism.

Furthermore, the equivalent carbon dioxide emissions for plant-based options are 10–50 times lower than for animal-based products (see Fig. 9.1) with the exception of dark chocolate and coffee which are discretionary foods and consumed in small quantities. Despite its increasing popularity, eating local may be good to

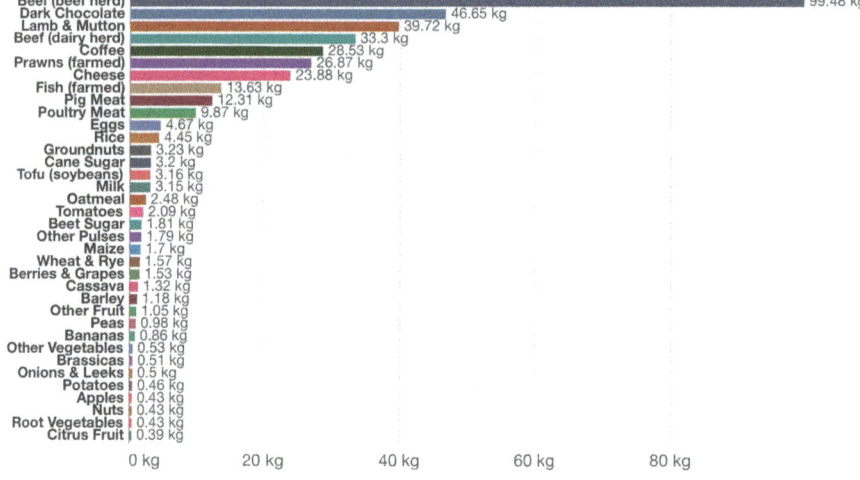

Greenhouse gas emissions per kilogram of food product

Greenhouse gas emissions are measured in kilograms of carbon dioxide equivalents (kgCO₂eq) per kilogram of food product. This means non-CO₂ greenhouse gases are included and weighted by their relative warming impact.

Source: Poore, J., & Nemecek, T. (2018). Reducing food's environmental impacts through producers and consumers. OurWorldInData.org/environmental-impacts-of-food · CC BY

Fig. 9.1 Greenhouse gas emissions of food. (Source: Ritchie & Roser, 2021)

socially and economically support communities, but it is not enough to counteract the factors fuelling the planetary emergency and is in fact misguiding (Ritchie, 2020). A reduction in the intake of animal-based products associated with large land use change and clearing of native vegetation should be the main focus to make human diets more sustainable.

The sustainability of human diets is linked to greenhouse gas emissions, use of land and other natural resources. Food production is uniquely impacting biodiversity in a way that no other human activities do. The sustainability of human diets was raised as an issue for the first time in the 1980s but because of the over-riding goal to feed a hungry and growing global population, it has been neglected for decades (Burlingame & Dernini, 2012). In 2010, the World Wildlife Fund–UK (Williamson, 2010) argued for a One Planet Diet which builds on the principles of quality food, social justice (including animal welfare), environment (including climate change, water, land use and biodiversity) and health (including safety and nutrition). A sustainable global diet should be "protective and respectful of biodiversity and ecosystems, culturally acceptable, accessible, economically fair and affordable; nutritionally adequate, safe and healthy; while optimising natural and human resources" contributing to nutritional security for present and future generations (Burlingame & Dernini, 2012, p. 7).

All evidence we have shows that human diets are far from sustainable and this has severe negative impacts. Between 1970 and 2016, global wildlife decreased by 68% (WWF, 2020) with a major cause being unsustainable agricultural practices associated with deforestation and clearing of native vegetation. In Australia some wildlife populations, including the iconic koalas, have plummeted by 97% and bio-diversity loss is happening at an unprecedented rate. More than 100 species have become extinct since colonialisation in 1788 and 1892 more species are at risk of extinction classified as "threatened" (Ecological Society of Australia, 2020). By shifting to more plant-based diets, we can significantly reduce extinction risks and "elephants, lions, hippos, orangutans, foxes, wolves, bears, even spiders – would have a much better chance of thriving in a world where humans eat less meat" (Greenpeace, 2018, p. 34).

Food waste is not considered part of any diet from a nutritional perspective. It is however highly relevant from a sustainability point of view. Approximately one third of food is lost from production to consumption (Gustavsson et al., 2011). This is a serious concern and there have been consistent calls to at least halve food waste (Willett et al., 2019), if not possible to eliminate it all together. Three aspects are of concern additionally. The first is very logical – when animal-based food is wasted, this has a higher ecological footprint than plant-based products. Such waste has been associated during its production with higher greenhouse gas emission, land use, water consumption and biodiversity impacts. To make things worse, livestock production also has the highest rate of agricultural losses with 81–94% of the global net primary product from croplands and grasslands lost (Alexander et al., 2017). The second is ethical – meat and other animal-based products represent inefficient use of resources through the longer chain of "feed–animal–food" instead of produc-ing directly food for human consumption (Schmidinger et al., 2018). In the case of beef, it takes 37 calories fed to the animal to produce one calorie for human con-sumption (Eshel et al., 2014). In other words, we are wasting resources through inefficient food production. Achieving better efficiency in food systems is required for global food security in a planet emergency. The third is related to overconsump-tion of food – when people are eating more than their nutritional requirements. Such excessive consumption is a loss in the food system (Alexander et al., 2017) resulting in products that could be eaten safely being wasted and misused causing obesity and overweight. It is often the case that such people also eat excessive animal-based products and do not consume enough whole grains, legumes, nuts and seeds (Willett et al., 2019). This eventually becomes a health problem and precondition for diabe-tes type 2, cardiovascular and other non-communicable diseases. Globally, there are more people who are obese and overweight – estimated at 1.9 billion (or 13.1% of the global population) in 2016 (WHO, 2020), than those who are undernourished – estimated at 658 million (or 8.8% of the global population) in 2016 and 688 million (8.9%) in 2019 (FAO, IFAD, UNICEF, WFP & WHO, 2020). The equity side of food consumption is also part of dealing with the planetary emergency. A COVID-19 induced food insecurity is also likely to have a more severe impact on the poorer people and countries as well as on women and the socially weak in society (FAO, IFAD, UNICEF, WFP & WHO, 2020). The assessment by Gustavsson et al. (2017)

shows that system losses from overconsumption of food across the globe, including livestock-based products, are at least as substantial as those from food discarded by consumers.

All three aspects of wasting food have serious sustainability implication. A shift towards flexitarianism seems a most logical way to achieve planetary health. It can also show that the western type of diet rich in animal-based foods is not ineluctable. However, those who consider themselves flexitarians need to not only reduce the intake of meat and other animal-based products but further embrace reduction and possible elimination of food waste as well as control over their own calorie intake. For example, the EAT-Lancet commission uses 2100 kcal per day per person as a reasonable average calorie intake (Willett et al., 2019).

Sustainable diets need to provide healthy levels of nutrition. In 2019, 26% of the world population experienced hunger or did not have regular access to nutritious and healthy food; around 144 million children under five were stunted for their age and a further 47 million pre-school children were wasted (Global Panel on Agriculture and Food Systems for Nutrition, 2020). Sub-Saharan Africa and South Asia continue to be the areas with the highest levels of food insecurity and with high rates of children's malnutrition, wasting and stunting (Roser and Ritchie, 2013). Food safety is another important aspect, particularly for places where there are less quality standards and monitoring. Without proper access to health and safe food, flexitarianism remains a chimera of the rich who do not need to worry about where the next meal will come from. In this respect, the Global Panel on Agriculture and Food Systems for Nutrition (2016, p. 24) talks about high-quality diets which "eliminate hunger, are safe, reduce all forms of malnutrition and promote health. High-quality diets are adequate, diverse and balanced. It is important that high-quality diets are produced without undermining the environmental basis for generating high-quality diets for future generations".

Malnutrition in this day and age is manifested through eating the wrong type and amount of food (e.g. obesity and overweight) and by not having access to healthy and nutritious choices (e.g. undernourishment and food insecurity). Although flexitarianism appears to be related directly to the former, it can seriously ease the burden on the latter by decreasing the competition for global resources, grains, land, clean air, fresh water and other environmental public goods on this planet. The decision every person makes with every meal consumed can shift the balance of our food systems in the wrong or right direction. This is a power we have as consumers but we should be making these choices in an environment that promotes good sustainable behaviour instead of having to fight cultural inertia, vested interests and misleading advices. Rather than flexitarians or vegaterians being singled out as following a different diet that somehow makes them unusual, special or poles apart from the rest of the affluent society, such a behaviour needs to simply become normal or mainstream.

Mainstreaming Flexitarianism

According to FAO, IFAD, UNICEF, WFP and WHO (2020, p. 49), "better diets are needed for everyone everywhere, regardless of cultural or religious patterns, or income at either national or household levels". There are immense benefits from flexitarianism that improve food security, nutrition and safeguard the health of the planet. The systems of producing food are changing but our choices of what to eat also need to evolve. In this planetary emergency, we are also facing a dietary crisis which is unlikely to be resolved by arguments along the lines which dietary pattern is the best or easiest to follow. Calamities, such as those related to climate change, COVID-19, social tensions and conflicts, are likely to exacerbate the consequences from our diet crisis. The vicious cycle between food and the planetary ecosystems needs to be broken with human diets holding the key to this conundrum.

There is ample evidence that under the current food systems, including waste, and western type of dietary preferences, the planet cannot provide for the entire human population. For example, a study by Gerten et al. (2020) estimates that only 3.4 billion people can be fed within the planetary boundaries. However, if cropland is redistributed away from producing animal feed, water–nutrient management is improved, food waste is reduced and diets change to a much lower animal protein intake (namely at 25% of the current levels in the respective food producing units), then we can feed comfortably 10 billion people (Gerten et al., 2020). Further lowering of the animal-based components in the human diet will allow to support even higher population numbers without relying on advances in science and technology, such as for example cultured meat. Another study by Theurl et al. (2020) of 520 different scenarios in a hypothetical no-deforestation world by 2050 shows that diets are the main determinant of greenhouse gas emissions. Diets with higher demand for ruminant meat and dairy generate the highest emissions and vegan diets the lowest (Theurl et al., 2020).

All this modelling indicates that flexitarianism or reduction in meat consumption needs to be normalised as essentially the human race has no other choice for a peaceful and gracious sustainable future. There have been previous attempts at introducing the idea about lower meat intake as a gradual process, such as nudging – interventions which encourage positive change without excluding undesirable options (Dagevos & Reinders, 2018), transitioning to more sustainable diets in a university college (Foenander et al., 2018) and initiatives, such as Meatfree Mondays (2021) supported by celebrities, such as Paul McCartney and Alicia Silverstone. Education can also play a role as can influence from friends (Bogueva et al., 2021b). These efforts however target the individual consumer who often feels that they have to swim against the flow. Instead of putting the onus on individual decision-making, the entire social environment should encourage flexitarianism. Below are some ideas as to how the broader community can be engaged in such a transition.

Working with the Medical Community

There is enough scientific evidence for doctors to encourage patients to maintain a healthy diet with reduced intake of animal-based products. The planetary co-benefits need to be part of the medical schools' curriculum and replace myths associated with red meat in particular (Bogueva & Phau, 2016). Dietary guidelines and specifically red meat limits by reputable health-based organisations need to be publicly displayed in hospitals, surgeries, doctors' rooms and waiting areas. Outdated understanding about the benefits of red meat needs to be put in context and replaced with the latest facts.

Start with the Children

In some cases, it may be easier to raise children in a low-meat diet rather than changing the habits of the entire family. Young children naturally opt for fruit, vegetables, nuts and seeds and not for meat-based options. This desire should be encouraged and maintained. Childcare centres and school canteens should similarly provide mainly plant-based fresh and nutritious food options.

Catering Services and Restaurants

Plant-based dishes should be the default option for any catering purposes with meat options available on request. This will be the opposite of the current situation where vegetarians and vegans need to put special dietary requests. In restaurant menus, again the range of options should be reversed with only a few meat-based options available and the majority of the dishes being plant-based. This similarly applies to fast-food restaurants and coffee shops.

Ban on the Advertising of Red Meat

In the same way we no longer market tobacco and cigarettes, advertisements promoting meat should not be allowed. Whilst cigarettes affect only the individual's health, meat exploits the global commons and affects personal and public well-being. There should be higher expectations about the duty of care by the government and public authorities. They should not allow the public marketing space to be used for unhealthy and unsustainable food choices.

Diversification of Sustainable Healthy Diets

The dominance of the Western meat-rich diet has homogenised food choices across the globe. This should be replaced with a variety of culturally appropriate, locally suited food options which stress taste and sustainability. Fish and seafood products can also form part of healthy diets but they need to be provided in a sustainable way that preserves the natural environment.

Fresh Produce Markets and Direct Delivery from the Farmer to the Consumer

Farmers' markets and local growers allow people to reconnect with the origins of the food they consume. The majority of the produce sold in such places is plant-based, in season and its freshness is attractive. Innovative systems for linking the farmers directly with the consumers using social media, blockchain and other information technologies can guarantee food provenance and quality.

Redirect Any Subsidies and Government Support to Plant-Based Options

Currently, the majority of government subsidies go to farmers who raise livestock (Shapiro, 2016). This is supporting businesses that would struggle otherwise and allowing them to remain profitable even under unfavourable climatic conditions. Such subsidies should be withdrawn and redirected to growers who produce plant-based foods.

Taxing Food to Internalise Costs to Society

Introducing taxes (e.g. value-added tax or goods and services tax) for foods who impose higher costs on society through burdening the public health system or negatively impacting the natural environment, is a logical way to reflect the costs imposed on society. The revenue collected through such taxes can be used to invest in research and development of better food alternatives as well as improvement of degraded agricultural land.

Transparency about Food Components and Labelling Reflecting the Sustainability of the Product

Consumers are used to health labels on food but they represent only one aspect of the appropriateness of dietary choices. A new labelling system can provide succinct information about the sustainability quality of a particular food product.

Improve the Taste Qualities of Plant-Based Alternatives

Many of the novel food alternative products imitate meat or processed meat. There is potential for innovation and new technological solutions to improve the current taste qualities of these novel food products. This can enrich the choices available to the consumers.

Reduction in Waste and Eliminating Plastics in Packaging

Attitudes towards food waste need to change and make it undesirable and unacceptable. Plastic packaging should be eliminated from all food products and replaced with plant-based and other environmentally friendly options.

The current clash between the food production systems and the ecological systems on which they depend should be seen as an opportunity for new and better things to emerge. They will leave not only better tastes in people's mouths but also a healthier planet.

Conclusion

According to Gustavsson et al. (2011), it is regrettable from environmental and food security perspectives to see projections that the consumption of meat, dairy and other animal-based products will continue to increase with rising incomes across the globe. Between 2000 and 2019, only a handful of countries have reduced their meat consumption (Whitton et al., 2021). Not only are there negative health implications, but the efficiency of the overall food system to provide for human nutrition is being lowered. Deliberate and conscientious abstaining from animal-based products is a reasonable answer to the myriad of negative implications associated with livestock which is significantly contributing towards the planetary emergency. In addition to climate change, biodiversity loss looms as a major anthropogenic failure originating from human diets.

Many policy reports link the issues related to food and nutrition to Sustainable Development Goal (SDG) 2 No Hunger. This is a good point of reference as SDG2 also includes specific targets that require policy frameworks and interventions to achieve. Diets, however, are a very personal behaviour and choice which are affected by many more factors outside of policy measures, including sensory experiences, psychological attitudes, health concerns, environmental considerations and marketing influences (Marinova & Bogueva, 2019). Although currently many people identify as flexitarians – 36% of the US consumers (Packaged Facts, 2020), only when flexitarianism is mainstreamed or normalised will the broader society transition to food choices that protect the environment and human health. The options listed here are a small selection of possible ways to make plant-based options the default preference in any society.

In some cases, this will be very easy to achieve, in others there may be opposition from those who are used to having meat on a daily basis. It is important in this transition to emphasise that the aim is not to eliminate meat and other animal-based foods altogether as they will still be available at a lower frequency. It is to replace them with food choices that are better, tastier and more sustainable. This will eventually lead to a moment in time when we will no longer require to use the term flexitarian and similarly flexitarianism will become absolute as all diets will be low on animal proteins protecting the health of people and the planet.

References

Alexander, P., Brown, C., Arneth, A., Finnigan, J., Moran, D., & Rounsevell, M. D. A. (2017). Losses, inefficiencies and waste in the global food system. *Agricultural Systems, 153*, 190–200. https://doi.org/10.1016/j.agsy.2017.01.014

American Dialect Society. (2004). *2003 words of the year*. https://www.americandialect.org/2003_words_of_the_year

Blatner, D. J. (2009). *The flexitarian diet: The mostly vegetarian way to lose weight, be healthier, prevent disease, and add years to your life*. McGraw Hill.

Bogueva, D., & Phau, P. (2016). Meat myths and marketing. In T. Raphaely & D. Marinova (Eds.), *Impact of meat consumption on health and environmental sustainability* (pp. 264–276). IGI Global.

Bogueva, D., Apostolova, M., & Danova, S. (2021a). Food, nutrition and health in Bulgaria. In A. Gostin, D. Bogueva, & V. Kakurinov (Eds.), *Nutritional and health aspects of food in the Balkans* (pp. 67–89). Elsevier.

Bogueva, D., Marinova, D., & Raphaely, T. (2021b). Influencing sustainable food-related behaviour changes: A case study in Sydney, Australia. In J. Bhattacharyya, M. K. Dash, C. Hewege, M. S. Balaji, & L. Weng (Eds.), *Social and sustainability marketing: A casebook for reaching your socially responsible consumers through marketing science (forthcoming)*. Routledge.

Bogueva, D., Marinova, D., & Todorov, V. (2021c). Bulgarian traditional folklore celebrating food and sustainability. *International Journal of Information Systems and Social Change, 12*(3), 1–14. https://doi.org/10.4018/IJISSC.2021070101

Burlingame, B., & Dernini, S. (Eds.). (2012). *Sustainable diets and biodiversity: Directions and solutions for policy, research and action*. FAO. http://www.fao.org/3/i3004e/i3004e00.htm

Clark, M., & Tilman, D. (2017). Comparative analysis of environmental impacts of agricultural production systems, agricultural input efficiency, and food choice. *Environmental Research Letters, 12*(6), 064016. https://doi.org/10.1088/1748-9326/aa6cd5

Dagevos, H., & Reinders, M. J. (2018). Flexitarianism and social marketing: Reflections on eating meat in moderation. In D. Bogueva, D. Marinova, & T. Raphaely (Eds.), *Handbook of research on social marketing and its influence on animal origin food product consumption* (pp. 105–120). IGI Global.

Derbyshire, E. J. (2017). Flexitarian diets and health: A review of the evidence-based literature. *Frontiers in Nutrition, 3*, 55. https://doi.org/10.3389/fnut.2016.00055

Doheny, K. (2020). *Mediterranean diet repeats as best overall of 2020.* https://www.webmd.com/diet/news/20200102/mediterranean-diet-repeats-as-best-overall-of-2020

Ecological Society of Australia. (2020). *Australia's species extinction crisis in numbers: 2019.* https://www.ecolsoc.org.au/?hottopic-entry=australias-species-extinction-crisis-in-numbers-2019

Eshel, G., Shepon, A., Makov, T., & Milo, R. (2014). Land, irrigation water, greenhouse gas, and reactive nitrogen burdens of meat, eggs, and dairy production in the United States. *Proceedings of the National Academy of Sciences of the United States of America (PNAS), 111*(33), 11996–12001. https://doi.org/10.1073/pnas.1402183111

European Commission. (2020). *Food-based dietary guidelines in Europe.* https://ec.europa.eu/jrc/en/health-knowledge-gateway/promotion-prevention/nutrition/food-based-dietary-guidelines

Foenander, E., Green, C., Portsmouth, L., & Raphaely, T. (2018). Marketing an environmentally sustainable catering model: A case study of Medley Hall Residential College in Victoria, Australia. In D. Bogueva, D. Marinova, & T. Raphaely (Eds.), *Handbook of research on social marketing and its influence on animal origin food product consumption* (pp. 267–282). IGI Global.

Food and Agriculture Organisation of the United States (FAO), International Fund for Agricultural Development (IFAD), United Nations Children's Fund (UNICEF), World Food Programme (WFP), & World Health Organisation (WHO). (2020). *The State of Food Security and Nutrition in the World 2020. Transforming food systems for affordable healthy diets.* Rome, FAO. http://www.fao.org/3/ca9692en/CA9692EN.pdf

Froggatt, A., & Wellesley, L. (2016). *China shows way with new diet guidelines on meat.* Chatham House. https://www.chathamhouse.org/2016/06/china-shows-way-new-diet-guidelines-meat

Gerten, D., Heck, V., Jägermeyr, J., Bodirsky, B. L., Fetzer, I., Jalava, M., … Schellnhuber, H. J. (2020). Feeding ten billion people is possible within four terrestrial planetary boundaries. *Nature Sustainability, 3*(3), 200–208. https://doi.org/10.1038/s41893-019-0465-1

Global Panel on Agriculture and Food Systems for Nutrition. (2016). *Food systems and diets: Facing the challenges of the 21st century.* https://www.glopan.org/foresight1/

Global Panel on Agriculture and Food Systems for Nutrition. (2020). *Future food systems: For people, our planet, and prosperity.* Foresight 2.0. https://www.glopan.org/foresight2/

Grant, J., & Richter, H. (2020). *2020: The year of the Flexitarian.* Sustainalytics. https://www.sustainalytics.com/esg-blog/2020-the-year-of-the-flexitarian/

Greenpeace. (2018). *Less is more: Reducing meat and dairy for a healthier life and planet.* The Greenpeace vision of the meat and dairy system towards 2050. https://www.greenpeace.org/static/planet4-international-stateless/2018/03/698c4c4a-summary_greenpeace-livestock-vision-towards-2050.pdf

Gustavsson, J., Cederberg, C., Sonesson, U., van Otterdijk, R., & Meybeck, A. (2011). *Global food losses and food waste: Extent, causes and prevention.* Food and Agriculture Organization of the United Nations (FAO).

Hirvonen, K., Bai, Y., Headey, D., & Masters, W. A. (2020). Affordability of the EAT–*Lancet* reference diet: A global analysis. *The Lancet Global Health, 8*, e59–e66. https://doi.org/10.1016/S2214-109X(19)30447-4

Italie, L. (2012). *New words make it into dictionary update.* https://www.news.com.au/world/breaking-news/new-words-make-it-into-dictionary-update/news-story/6cff131499089aa5ab996b8bd16f67dc

Kostova, S., Atanasov, B., & Marinova, D. (2018). Consumption of animal products in Bulgaria: The case for change. In D. Bogueva, D. Marinova, & T. Raphaely (Eds.), *Handbook of research on social marketing and its influence on animal origin food product consumption* (pp. 283–297). IGI Global.

Marinova, D., & Bogueva, D. (2019). Planetary health and reduction in meat consumption. *Sustainable Earth, 2*, 3. https://doi.org/10.1186/s42055-019-0010-0

Meatfree Mondays. (2021). *One day a week can make a world of difference.* https://www.meatfreemondays.com

Mintel. (2021). *2021: Global food and drink trends.* https://downloads.mintel.com/private/jcZU4/files/852695/

National Health and Medical Research Council (NHMRC). (2013). *Eat for health: Australian National Dietary Guidelines.* Australian Government. https://www.eatforhealth.gov.au/sites/default/files/content/n55_australian_dietary_guidelines.pdf

Packaged Facts. (2020). *Vegan, vegetarian, and flexitarian consumers.* https://www.packagedfacts.com/Vegan-Vegetarian-Flexitarian-Consumers-13656739/

Poore, J., & Nemecek, T. (2018). Reducing food's environmental impacts through producers and consumers. *Science, 360*(6392), 987–992. https://doi.org/10.1126/science.aaq0216

Raphaely, T., Marinova, D., Crisp, G., & Panayotov, J. (2013). Flexitarianism (flexible or part-time vegetarianism): A user-based dietary choice for improving personal, population and planetary wellbeing. *International Journal of User-Driven Healthcare, 3*(3), 34–58. https://doi.org/10.4018/ijudh.2013070104

Roser, M., & Ritchie, H. (2013). *Hunger and undernourishment.* Our World in Data. https://ourworldindata.org/hunger-and-undernourishment#too-little-height-for-age-stunting

Ritchie, H., & Roser, M. (2019). *Meat and dairy production.* Our World in Data. https://ourworldindata.org/meat-production

Ritchie, H., & Roser, M. (2021). *Environmental impacts of food production.* Our World in Data. https://ourworldindata.org/environmental-impacts-of-food

Schmidinger, K., Bogueva, D., & Marinova, D. (2018). New meat without livestock. In D. Bogueva, D. Marinova, & T. Raphaely (Eds.), *Handbook of research on social marketing and its influence on animal origin food product consumption* (pp. 344–361). IGI Global.

Shapiro, P. (2016). Feasting from the federal trough: How the meat, egg and dairy industries gorge on taxpayer dollars while fighting modest rules. In T. Raphaely & D. Marinova (Eds.), *Impact of meat consumption on health and environmental sustainability* (pp. 244–254). IGI Global.

Sparkes, D. (2019). *Australian Heart Foundation changes guidelines for egg and dairy consumption.* ABC The World Today. https://www.abc.net.au/radio/programs/worldtoday/heart-foundation-changes-guidelines-egg-and-dairy-consumption/11435004

Starostinetskaya, A. (2020). *Interest in veganism hits all-time high in 2020, Google trends report shows.* https://vegnews.com/2020/9/interest-in-veganism-hits-all-time-high-in-2020-google-trends-report-shows

Streit, L. (2019). *The flexitarian diet: A detailed beginner's guide.* https://www.healthline.com/nutrition/flexitarian-diet-guide

Taub-Dix, B. (2019). *What is a flexitarian diet? What to eat and how to follow the plan.* https://www.everydayhealth.com/diet-nutrition/diet/flexitarian-diet-health-benefits-food-list-sample-menu-more/

Theurl, M. C., Lauk, C., Kalt, G., Mayer, A., Kaltenegger, K., Morais, T. G., … Haberl, H. (2020). Food systems in a zero-deforestation world: Dietary change is more important than intensification for climate targets in 2050. *Science of the Total Environment, 15*, 735. https://doi.org/10.1016/j.scitotenv.2020.139353

US News Staff. (2021). *U.S. News best diets: How we rated 39 eating plans.* https://health.usnews.com/wellness/food/articles/how-us-news-ranks-best-diets

Vitarama. (2020). *What is Christmas Lent and how to prepare for it?* https://vitarama.bg/bg/kakvo-sa-kolednite-posti-i-kak-da-se-podgotvim/401/item/ (in Bulgarian).

Whitton, C., Bogueva, D., Marinova, D., & Phillips, C. J. C. (2021). Are we approaching peak meat consumption? Analysis of meat consumption from 2000–2019 in 35 countries and its relationship to Gross Domestic Product. *Animals, 11*, 3466, https://doi.org/10.3390/ani11123466

Willett, W., Rockström, J., Loken, B., Springmann, M., Lang, T., Vermeulen, S., … Murray, C. J. L. (2019). Food in the Anthropocene: The EAT–Lancet Commission on healthy diets from sustainable food systems. *The Lancet, 393*(10170), 447–492. https://doi.org/10.1016/S0140-6736(18)31788-4

Williamson, D. (2010). *WWF-UK policy position statement on food.* http://assets.wwf.org.uk/downloads/food_position_statement_final_april_2010.pdf

World Cancer Research Fund (WCRF) & the American Institute for Cancer Research (AICR). (2018). *Recommendations and public health and policy implications.* https://www.wcrf.org/sites/default/files/Recommendations.pdf

World Health Organisation (WHO). (2020). *Obesity and overweight.* https://www.who.int/news-room/fact-sheets/detail/obesity-and-overweight

World Wildlife Fund (WWF). (2020). *Living Planet Report 2020: Bending the curve of biodiversity loss* (R. E. A. Almond, M. Grooten & T. Petersen, Eds.). WWF.

Chapter 10
Mitigating Diseases

Abstract It appears that a disconnect between food and disease mitigation has occurred with development. In contrast, the chapter examines the link between food and non-communicable diseases, then the threats caused to human health by zoonotic viruses and infections, and finally the prospects of losing the power of antibiotics to improve human conditions. A major contributing factor to these health-related conditions and threats is the excessive consumption of meat and other animal-based products. Replacing them with plant-rich food options can have numerous benefits.

The original version of the Hippocratic Oath states: "I will use those dietary regimens which will benefit my patients according to my greatest ability and judgment, and I will do no harm or injustice to them" (North, 2002, para. 5). This 2500-year-old way of expressing the doctors' commitment to their patients, the profession and humanity has seen updates to reflect the spirit of modern times and the Declaration of Geneva represents this (AMA, 2006). One of the changes that have occurred is that there is no longer reference to diets in the vow medical practitioners around the globe take.

This is indicative of the disconnect between food and disease mitigation that has occurred with development. The lack of commitment to advice on healthy diets is not intended to be critical or diminishing of the excellent work doctors and the medical profession do, particularly in complex and difficult situations such as the COVID-19 pandemic. However, the disengagement with food has become persistent. Instead of influencing people's dietary behaviours, doctors are looking for other technical solutions, e.g. through medication or surgical interventions. The increasing prevalence of non-communicable diseases combined with the current and emergent threats of zoonotic outbreaks, epidemics or pandemics, is to a large degree a consequence from the disconnect between food and health that is happening within contemporary society.

In this chapter, we first examine the link between food and non-communicable diseases, then the threats caused to human health by zoonotic viruses and infections, and finally the prospects of losing the power of antibiotics to improve human

D. Marinova, D. Bogueva, *Food in a Planetary Emergency*,
https://doi.org/10.1007/978-981-16-7707-6_10

conditions. A contributing factor to these health-related conditions and threats is the excessive consumption of meat and other animal-based products. At present, global food choices are threatening both human health and the ecological state of the planet, to the extent that unhealthy diets pose a greater risk to morbidity and mortality than do unsafe sex, alcohol, drug and tobacco use combined (Willett et al., 2019). Let's unpack what this means for human health.

Non-communicable Diseases

What people eat has direct consequences for their health. Good dietary choices result in enhanced physical and mental wellbeing while also putting less pressure on the natural environment and its biophysical systems. For a long time, food has been seen as an issue outside the policy agenda related to climate change and sustainability because of two reasons. In the wealthier countries where many food options are freely available, what one eats is considered an issue of individual freedom of choice. On the other hand, in many places around the world starvation and food shortages continue to be a real threat to human survival. Despite the human race producing enough food to feed all people, uneven distribution and misappropriation are impacting its availability. Globally, there are now 2 billion people who are overweight or obese and 820 million who are hungry (FAO, IFAD, UNICEF, WFP & WHO, 2019). The gaps in food security exist not only between different parts of the world but also within each country with the socially weak groups of the population being exposed to a greater risk of malnutrition and poor dietary choices. With malnutrition and undernourishment existing and in fact, expanding, it is very difficult to achieve the UN Sustainable Development Goal 2 Zero Hunger and its targets. What is even more worrying is the direct links between overexploitation of nature's resources to produce food and the escalation of the public health problems for those who exceed a healthy body mass.

Prior to COVID-19, non-communicable diseases (NCDs) accounted for 71% of all deaths globally killing yearly 41 million people (WHO, 2018). This was and is likely to continue to be the most important public health challenge of the twenty-first century in terms of the global burden of disease and disability caused by unhealthy diets, physical inactivity as well as tobacco and alcohol consumption. The burden of NCDs, such as diabetes mellitus type 2 (also referred to as type 2 diabetes), heart disease, cancer and dementia, are all linked to our dietary choices with obesity and overweight being a major risk factor. Obesity is also correlated with a higher probability for positive COVID-19 test results (Yang et al., 2020) but also with the related severity and mortality outcomes from the disease (Chu et al., 2020; Zhao et al., 2020). There is ample evidence that plant-based dietary options and reduced meat intake decrease the prevalence of NCDs and obesity (Marshall & Marinova, 2019). Although the mechanisms why this is the case are not always clear, the empirical evidence is convincing. Many of the NCDs are also chronic which means that they are lasting conditions with persistent effects and people have been exposed for prolonged periods of time to the factors that trigger them. Below we examine the link between specific NCDs and food choices, starting with obesity.

Obesity

In a planetary emergency when the planet is literally burning and we are losing native species at an unprecedented rate, how much food people consume becomes an issue of social justice not only to the current, but also to the future generations. There are of course medical conditions (such as underactive thyroid, Cushing's syndrome or polycystic ovary syndrome) or intake of medicaments which result in low metabolism and may cause weight gain; however, our concern is about the mass of people who are overweight or obese because of behavioural and social factors. In the latter case, there is an imbalance between energy intake from food and energy use by the human body (AIHW, 2020). The increasing rates of obesity around the globe raise concerns not only about public health but also that the global commons are being overused inefficiently and ineffectively. With the environmental footprint of animal-based foods being generally higher than that of plant-based foodstuffs (Ritchie & Roser, 2020) and when this excess energy is supplied by livestock products, people's dietary behaviours further exacerbate the climate change and environmental emergency.

Red meat, with beef in particular, is the most detrimental food option from a planetary health perspective. This was confirmed by a meta-analysis of 570 studies from 119 countries which examined the life-cycle environmental footprints of 40 different food items that provide 90% of the global protein and calorie intake (Poore & Nemecek, 2018). An important dimension of the planetary emergency associated with food is eutrophication of the water bodies and ecosystems because of excessive levels of nutrients which enter the environment through agricultural practices and particularly the use of fertilisers (Ritchie & Roser, 2020). According to the eutrophying emissions per one kilogram of food produces, beef, followed by farmed fish and prawns, by far exceed these of plant-based options (see Fig. 10.1). Hence, the excessive intake of animal-based products is not only contributing to obesity but also destroying the foundations of producing food on this planet.

Furthermore, good gut health is an important factor that facilitates the digestive processes influencing metabolism in the body and immune responses to inflammation. The analysis by Dahl et al. (2017) shows that fibre intake through fruit, vegetables, legumes, wholegrains and nuts, has a clear-cut positive effect on the gut microbiota and lowers the risk of chronic diseases. High intake of red meat on the other hand is associated with obesity and increased waist circumference (Ekmekcioglu et al., 2018) that predispose to chronic diseases and multimorbidity (that is simultaneous presence of several NCDs). In all respects, diets which involve many plant-based foods, such as the Mediterranean, score better from a public and environmental health perspective in the current planetary emergency. They are also associated with lower levels of obesity and overweight (Tuso et al., 2013; Medawar et al., 2019; Brazier, 2020).

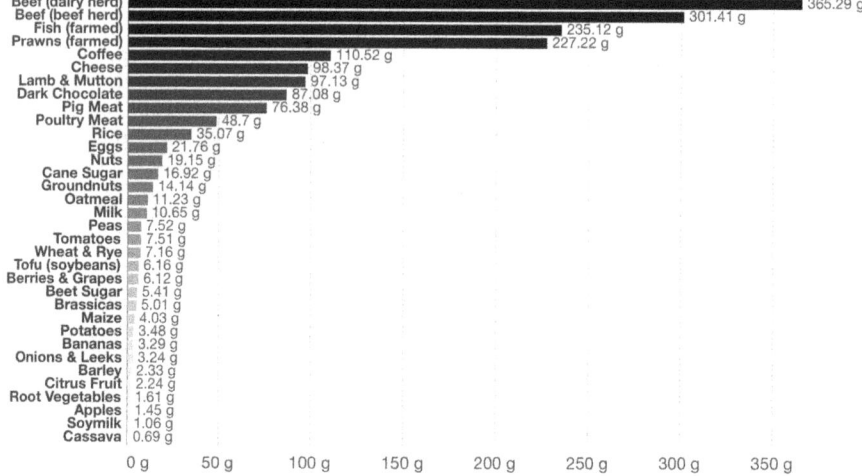

Fig. 10.1 Eutrophying emissions per one kilogram of food. (Source: Our World in data, https://ourworldindata.org/environmental-impacts-of-food)

Cardiovascular Diseases

Cardiovascular diseases (including coronary heart disease, heart failure and stroke) remain the leading cause of death in the world even during the COVID-19 pandemic. They caused the deaths of almost 18.6 million people globally and experts expect that the long-term effects of COVID-19 will further increase the burden of this disease (American Heart Association, 2021). Many of these serious conditions can be prevented by addressing behavioural risk factors and improved diets (WHO, 2017a).

A review of the health implications of a vegetarian diet presents significant evidence that diets with a high intake of wholegrains, legumes and nuts and low in meat, such as in the Mediterranean diet, are associated with lower risk of cardiovascular diseases (Marsh et al., 2012). Haddad (2018) explains that these benefits come from both, the foods consumed by vegetarians and those avoided by them, such as meat. Another review of red meat, diseases and healthier alternatives (Ekmekcioglu et al., 2018) showed nuts and legumes lowering the risk of ischemic heart disease while a meta-analysis of fruits and vegetables consumption and stroke provides evidence of an inverse relationship between intake and risk of stroke (Hu et al., 2014).

Contributing factors for cardiovascular diseases are: higher intake of sodium (Aburto et al., 2013b) more common with the consumption of animal proteins; and lower intake of potassium (Aburto et al., 2013a) present in unrefined plant-based foods, including beans, chickpeas, peas, lentils, nuts as well as leafy green vegetables such as spinach, cabbage, kale and parsley, and fruits such as berries, bananas, dates, paw-paw and pomegranates. For example, the Dietary Approaches to Stop Hypertension (or DASH) diet designed as a healthy eating plan to prevent high blood pressure is low on sodium and rich in vegetables, fruits and wholegrains.

The environmental benefits from all plant-based options are indisputable compared to animal-based foods (Aleksandrowicz et al., 2016; Springmann et al., 2016; Clark & Tilman, 2017; Shepon et al., 2018). However, from a health point of view, only healthy choices, such as wholegrains, fruits, vegetables, nuts, legumes, oils, tea and coffee, are associated with a lower risk of cardiovascular diseases (Satija et al., 2017; Wei et al., 2016).

Type 2 Diabetes

More plants and less meat is also the message from studies that analyse the impacts of diets and foods on insulin levels and managing of type 2 diabetes. Wholegrains are particularly beneficial as they have a lower glycaemic index and less impact on blood glucose (Ekmekcioglu et al., 2018) contributing to a reduced risk of diabetes (Barclay et al., 2008; Greenwood et al., 2013; Bhupathiraju et al., 2014). A review of red meat, diseases and healthier alternatives found that legumes are similarly very good for lowering blood glucose and glycosylated haemoglobin, increasing insulin sensitivity (Ekmekcioglu et al., 2018). Furthermore, there is evidence that animal proteins themselves are associated with an increased risk of type 2 diabetes (Malik et al., 2016; Shang et al., 2016; Marsh et al., 2016). Replacing red meat with vegetable protein improves kidney function which reduces the risk and progression of renal disease in people with any type of diabetes (Marsh et al., 2012).

Despite such clear evidence, the consumption of meat and other animal proteins continues to be very high in countries, such as Australia, Brazil and USA, and it is likely that the cases of type 2 diabetes will continue in the future. Co-morbidities resulting from diabetes include chronic kidney disease, stroke and cardiovascular disease. The situation is particularly worrying because the world is increasingly witnessing higher numbers of children and young adults with type 2 diabetes (Nadeau et al., 2016). Their food choices should be directed towards healthier options which co-incidentally are also environmentally better.

Cancer

The world's largest and most reliable scientific study on cancer prevention is the Continuous Update Project (WCRF/AICR, 2018). This is an outgoing programme in which the world's best experts continuously update dietary and nutritional recommendations to help cancer prevention and survival. Their Third Expert Report (WCRF/AICR, 2018) recommends eating wholegrains, non-starchy vegetables, fruit and legumes because they have a protective effect against cancer, particularly colorectal and aerodigestive cancers. On the other hand, the intake of red and processed meat should be limited. The digestive tract is directly exposed to the food we consume and plant-based healthy options, such as wholegrains and legumes, protect from gastrointestinal cancers (Ekmekcioglu et al., 2018). There is also evidence that consuming red and processed meat increases the risk of other cancers, such as lung and pancreatic cancer (WCRF/AICR, 2018).

Based on consistent evidence from 800 studies, the World Health Organization (2015) categorised processed meats as Group 1 carcinogenic to humans (together with tobacco and asbestos) and red meat as Group 2A probably carcinogenic (together with the herbicide glyphosate). Systematic literature reviews consistently confirm the benefits of diets based on low meat intake (e.g. Contreras García & Zaragoza-Martí, 2020). These categorical messages are yet to be properly accepted by the wider community in wealthier countries or wealthier sections of society in developing countries. It is very important to use social marketing (Bogueva et al., 2021) and other methods to once again communicate that plant-based foods that have less impact on the natural environment are also better and protective for human health.

If a single food item, such as meat can change the burden of disease so profoundly with massive public health implications, it is worth spreading the knowledge and encourage humanity to pursue dietary changes. Many diets which emphasise the intake of vegetables are known to be healthier and protective against cancer while having a much lighter environmental footprint. Examples include the Mediterranean diet – rich in grains, fruit, vegetables, herbs, spices, nuts, beans, healthy oils with a limited intake of dairy and seafood, and globally recognised as protective against heart attack, as well as the Okinawa diet in Japan – low in calories and fat content and high in carbohydrates, encouraging the intake of vegetables and soy products with small amounts of fish and pork, linked to the high longevity of the people on this island and their lower cancer rates. As we all differ in our consumption patterns and food preferences, we do not to strictly adhere to a particular diet everywhere across the world, but the underlying principles are globally valid. By embracing more vegetables, legumes and fruits and reducing the intake of animal-based products, we can adjust our diet to be culturally and contextually appropriate.

The low intake of vegetables and fruit across the world, both in higher- and lower-income countries, results in many avoidable deaths globally (Global Panel on Agriculture and Food Systems for Nutrition, 2016), estimated at 3.9 million in 2017 (WHO, 2019). In India for example, the per capita consumption of fruit and vegetables is 130 grams per day as opposed to the recommended by the World Health Organisation intake of 400–600 grams per day (Mason-D'Croz et al., 2019; ICMR & NIN, 2019) or the 550 grams per day according to the EAT-Lancet Commission Planetary Diet (Willett et al., 2019). The importance of fruits and vegetables for human nutrition, health, food security and the achievement of the Sustainable Development Goals (SDGs) is reflected in the United Nations (UN) General

Assembly pronouncing 2021 as the International Year of Fruits and Vegetables (WHO, 2021a).

Mental Health

Increased intake of fruit, vegetables, legumes and wholegrains has many co-benefits, including better mental health and reduced risk of cognitive impairment and dementia. This is another win-win situation that is not only better for the natural environment but also to people's psychological health. A British study of young adults shows that a daily intake of 7–8 servings of vegetables and fruit results in positive emotional experiences (White et al., 2013). Similar results are reported in Chile (Piqueras et al., 2011), South Korea (Kye & Park, 2014) and USA (Boehm et al., 2013). According to Blanchflower et al. (2012, p. 1), "happiness and mental health rise in an approximately dose-response way with the number of daily portions of fruit and vegetables. The pattern is remarkably robust to adjustment for a large number of other demographic, social and economic variables". By reducing the consumption of animal-based foods and increasing the healthy plant-based options, we are literally feeling better while supporting the health of the planet.

Numerous studies confirm the benefits of diets which contain more plant-based foods and less meat, such as the Mediterranean diet, for cognitive function and protection against dementia (Petersson & Philippou, 2016). There are similarly confirmed benefits for decreasing the risk of Alzheimer's disease (Otaegui-Arrazola et al., 2014). With population ageing, brain health becomes even more important and the benefits of such diets are even more pronounced (e.g. Rajaram et al., 2019). High intake of vegetables in particular is associated with slower cognitive decline and a lower risk of dementia in older age (Loef & Walach, 2012). A diet high in plant-based foods, especially green leafy vegetables, and low in animal and high saturated fats is similarly recommended to prevent chronic diseases which also impact mental health (Ross, 2018). The answer for better mental health seems to be more plants and less meat.

Zoonotic Diseases

Zoonotic diseases (or zoonoses) are those caused by harmful microorganisms, such as viruses, bacteria, fungi and parasites, which originate from animals, including insects, and spread to people causing illness, discomfort and in some cases death (CDC, 2017). Vector-borne diseases, such as malaria (caused by a protozoan parasite transmitted by mosquitoes), dengue fever (four types of dengue virus also transmitted by infected *Aedes aegypti* type mosquitoes found in tropical and sub-tropical areas), Ross River virus (another mosquito-transmitted infection) or Lyme disease (a tick-borne infection with bacteria from the *Borrelia burgdorferi sensu lato* group), can also be considered zoonotic (Ramasamy, 2014). They are transmitted to the person through an insect "vector". In some cases of viruses that have been around for a relatively long time and have had high impacts, such as yellow fever (also transmitted by mosquitoes), rabies (transmitted through human exposure to the saliva or nervous system/brain tissue of an infected mammal animal which can occur through a bite, a person's broken skin or eyes, nose and mouth) and in the past, plague (transmitted through bites of infected fleas), there are vaccines that can prevent infection; while in others inoculations are yet to be developed. Snakebites are similarly zoonotic, highly impactful diseases for which antivenoms are administered to boost the sufferer's immune system and produce antibodies.

The majority of pathogens that cause infection in humans, between 58% and 75%, originate from microorganisms hosted in non-human animals (Ellwanger et al., 2021). However, the World Health Organisation (2020) classifies as zoonotic only infections originally transmitted from vertebrate animals to humans which then can be passed from human to human. This is the definition that we adopt when examining the link between food and zoonotic diseases. This implies that we exclude potential diseases that can originate when insects are used as food, the main reason for this being that such options are not yet mainstream in most parts of the world. We are also not interested in bioterrorism viruses nor in pathogens that can develop during cooking, preserving or storing of food. Even with these limitations, the group of zoonotic diseases is quite large and covers a range of conditions associated with human contact and interactions with the animal world, but our focus is only on food-producing animals or livestock.

In order to transmit the zoonotic infection, the farm animals themselves are infected first and often exhibit symptoms and conditions similar to the human illnesses. Bacteria and viruses affecting food-producing animals have always existed in the air, soil, water and wildlife, including predecessors of livestock. However, they thrive better in dense environments with numerous possible hosts than in the wild. Human ingenuity has been looking to find ways to deal with bacteria and viruses, including vaccination, prevention and treatment with drugs, such as antimicrobials, isolation and killing of sick animals and preventative culling of entire herds and flocks when the risk for human transmission becomes too high. This has been the case with the food-and-mouth disease (whose virus has seven serotypes – A, O, C, SAT1, SAT2, SAT3 and Asia1, and more than 60 different strains but does pose a serious threat to humans), bovine spongiform encephalopathy (or mad cow disease due to infection with a misfolded protein, known as a prion, which can spread to humans as a variant of the Creutzfeldt-Jakob neurodegenerative disease) and avian influenza (H5 and H7 subtypes of avian influenza A viruses are considered highly pathogenic to humans) which were related respectively to cloven-hoofed animals, such as cattle, sheep and pigs, cattle and poultry livestock (te Beest et al., 2011). A similar culling response was put in place in reaction to the classical swine fever (or hog cholera caused by the small RNA *Pestivirus* virus from the family *Flaviviridae*) which does not affect people. In 2019, the outbreak and spread of African swine fever (a large DNA virus from the *Asfarviridae* family unrelated to the classical swine fever), also considered harmless to humans, similarly resulted in massive culls of millions of pigs in China, Vietnam and across Asia. The environmental consequences in terms of soil, air and water pollution and contamination from the dead animal bodies as well as from the application of disinfectants and other chemical cleaning of premises are not estimated and are seen only as economic losses. Often it is even difficult to maintain the moral and ethical grounds about the carnage, destruction and devastation associated with livestock culling.

Driven by the constantly raising demand for animal-based food products, intensified livestock production occurs in inadequate living conditions with overcrowding and large flocks, and represents the biggest threat to the spread of infections. Hence, people's food preferences are ultimately a driver of zoonotic disease emergence, transmission and infection (National Research Council, 2009). A solution put forward to resolve the conflict between humans and livestock when it comes to zoonotic diseases is the One Health approach which acknowledges and accepts the links between people, animals and the natural environment (Stuchin et al., 2016). It recognises that "human and non-human animal health is interlinked through our shared environment" and also aims at improving the condition of the ecological systems (Degeling et al., 2016, p. 244). Under the One Health approach, health is reframed not only as a public good, but as a universal good shared across species. In other words, the response to zoonotic risks should be policies and practices that "explicitly seek to co-promote human and non-human benefits" (Degeling et al., 2016, p. 250). For example, instead of culling farm animals under the fear of contamination, the focus should shift from the specific pathogens to the drivers that facilitate the manifestation of the zoonotic disease as a social, economic, medical and environmental threat (Hinchcliffe, 2015). At the core of these drivers is the current and raising demand for animal-based foods and unless this is curbed, the One Health approach will remain a high moral stance or epistemic watchword (Michalon, 2020) inconsistent with the reality on the ground.

Back in 2003, the US Institute of Medicine (2003, p. 19) warned about the emerging new microbial threats to public health: "A transcendent moment nears upon the world for a microbial perfect storm… the microbial perfect storm will be a recurrent event". A combination of factors drives the emergence and spread of new zoonotic diseases and includes among others the globalised trade system and international travel. In fact, zoonotic diseases represented 75% of all emerging pathogens during the last decade (WHO, 2021b). Livestock production, steered by the human demand for animal protein, is also identified as a driving factor of the perfect storm as these animals are potential reservoirs of pathogens epidemiologically connected with human populations (Haydon et al., 2002; WHO, FAO & World Organisation for Animal Health, 2004). Let's examine some of the impacts of this factor on the emergence of new zoonotic diseases:

- livestock numbers – there have been endless discussions and commentaries about the growing global human population which in 2018 reached 7.6 billion (Population Reference Bureau, 2018). What receives significantly less attention is that people are vastly outnumbered by livestock animals. In that same year – 2018, there were 72 billion farm animals slaughtered for meat, including chicken, pigs, turkey, sheep, goats and cattle (Our World in Data, n.d.). In addition, there were 1.5 billion cattle alive (FAOSTAT, 2021). These animals are often raised in confined conditions in close proximity to one another where they cannot exhibit their natural behaviours. Such farm animals are not only severely stressed but are also large potential reservoirs for pathogens. Mammals species in particular harbour more pathogens (Gibb et al., 2020) but the world has already also witnessed significant zoonotic outbreaks originating from poultry, such as the avian influenza.
- shared environment – humans and farm animals are frequently in contact as they share the same surroundings and this facilitates transmission of pathogens. Contact between people and animals often occurs directly through handling or through feeding, water or simply sharing the same environment. Intensifying livestock husbandry to deliver more animal-based foods has clear epidemiological consequences with higher levels of risk of zoonotic diseases (National Research Council, 2009). Frequent and intense contacts between humans and farm animals "increase the chance of spillover events between these species and the emergence of new infectious diseases in humans. If the pathogen that was introduced into the human population is faced with favourable conditions for its replication and spreading, outbreaks or epidemics may occur" (Ellwanger et al., 2021, p. 4). Physical proximity to animals and animal density facilitate such transmissions.
- land clearing, deforestation and encroaching into areas with native vegetation – the land requirements of livestock for grazing and feed production (mainly based on exotic species) are estimated to currently take up 27% of all available landmass on this planet[1] (Ritchie & Roser, 2019). This has resulted in biodiversity loss and degradation of essential ecosystems which were originally covered with native vegetation, such as forests, steppes, savannas and rangelands. Biodiverse natural environments are more resilient and less susceptible to pathogens. Shifts in land use with increased human presence also change the host-pathogen interactions and exposure to vector dynamics. An Australian study, for example, concludes that there is an association between land clearing and accelerated infectious disease emergence (McFarlane et al., 2013). A global analysis of 6801 ecological settings and 376 host species shows that land use change significantly affects local host zoonotic communities creating expanding hazardous interfaces between livestock and people (Gibb et al., 2020).

---

[1] By comparison, the landmass area used as cropland for all other crops, except animal feed, takes up 7% of the world's land (Ritchie & Roser, 2019).

- climate change – because of temperature changes, global warming is shifting the areas where vector-borne diseases are present and is also changing the conditions for producing food. Livestock is substantially contributing to greenhouse gas emissions and deforestation which affect the global climate and water cycles. Food-producing animals are also particularly vulnerable. Extreme weather events, such as heat waves, heavy rains, floods and fires, as well as periods of limited supply of feed or water, impact negatively on livestock. Even when they physically survive the weather calamity, farm animals are still exposed to suffering and strain. Stressed animals have weaker immune systems and are more susceptible to harbouring pathogens which their organisms cannot fight (Phillips, 2020). This makes livestock animals potential reservoirs of pathogens for human transmission and infection.

We are writing this book during COVID-19 – the coronavirus pandemic caused by the severe acute respiratory syndrome coronavirus 2 (SARS-CoV-2). This SARS virus is the second severe and easily transmissible disease to emerge in the twenty-first century. The first was in 2002 and caused by a SARS virus whose precursor was traced to a wild animal, possibly palm civet or racoon dog (Poon et al., 2004). Deemed inefficient to directly infect people, the transfer of the coronavirus to humans most likely occurred after amplification in environments, such as live markets, where there is a diversity of mammal species (Poon et al., 2004). The origin of the SARS-CoV-2 virus is also probably zoonotic, possibly linked to a horseshoe bat (Mackenzie & Smith, 2020). Although this is not fully confirmed (Haider et al., 2020), virologists agree that despite the natural reservoir of SARS-CoV-2 being particular wildlife species, similarly to SARS, the virus needed another intermediate host before it reached humans (Spinney, 2020). This intermediate animal and the origin of SARS-CoV-2 may never be identified but the pandemic brings to the fore fundamental issues about how humans interact with the natural environment.

Preserving natural habitats obviously reduces the risk of diseases spilling over from wild animals (Burki, 2020). Furthermore, the analysis by Gibb et al. (2020) shows that as wild areas become agricultural land vastly dominated by livestock production, there is greater abundance and broader variety of wildlife species that carry pathogens and parasites capable of infecting humans. The species that are more sensitive to human impacts are lost from these landscapes while those who survive the land-use change are more resilient and "are more likely to carry disease" (Gibb quoted in Mukpo, 2020, para. 12). The surviving wildlife species are exposed to environmental stress and thus more likely to expel large doses of virus increasing the risk of an epidemic among the host population and of spillover to people (Burki, 2020). By allowing livestock agriculture to continue to perturb and convert wild areas across the globe, we are making all corners of the world susceptible to increasing risks of infection diseases. A way to control this and prevent further emergence and spread of new zoonotic diseases is to stop land conversion, deforestation, habitat destruction and disturbing the natural environment by containing agriculture within its current boundaries. This can only be achieved if we also put a stop on the increasing demand for animal-based foods.

In addition, the experience with COVID-19 shows that many meat-processing and packaging facilities, including slaughter houses, became hotspots in disseminating the infection because of the working conditions with cold temperatures and metallic surfaces that allow longer survival of the virus, and physical proximity between workers (Middleton et al., 2020). In Australia, the high risk in the meat processing industry is acknowledged with people who work there being given a priority in receiving a COVID-19 vaccine (together with defence, police, fire and emergency services). Meat proves to carry overall a greater health risk than plant-based options and calls have been made that the food industry should be held responsible not only for obesity and related non-communicable diseases but also for COVID-19 (Tan et al., 2020; Medawar et al., 2019).

Antimicrobial Resistance

In the last decade or so, doctors have become quite careful when prescribing antibiotics to patients as they are aware of how valuable these substances are for treatment and prevention of human bacterial and microbial infections. Since the boom in medical discoveries between the 1940s and 1980s, only one new class of antibiotics (namely, Oxazolidinones) has been added to the list of these precious substances with the remaining entrants being variations of existing antimicrobials (Sengupta et al., 2013). Many researchers and doctors warn that the "clinical management of antibiotic-resistant superbugs is assuming the shape of a gruesome problem because of the depleting antibiotic reserve" (Sengupta et al., 2013, p. 6). In fact, resistance has been observed to almost all antibiotics that we currently have (Ventola, 2015), a major contributing factor to which is overuse. There is ample evidence from epidemiological studies about the direct link between overuse of antibiotics through over-prescription and the emergence and dissemination of resistant bacteria strains (Editorial, 2013). After many decades of successful use of antibiotics, we are facing the threat of losing their beneficial effects for treatment and prevention of diseases and medical conditions in medicine and surgery. Numerous calls are being made to change the medical paradigm and use antibiotics as a last resort when other therapeutic means have failed (Michael et al., 2014).

The looming crisis in antimicrobial resistance is further massively exacerbated by the use of antibiotics for livestock. With the intensification of this industry since the 1950s, the farm animals are increasingly being raised concentrated in restricted areas, particularly evident in the case of poultry, pigs, aquaculture fish and feedlot systems for cattle and sheep. In USA, 80% of all sold antibiotics are given to animals, mainly to promote growth and to prevent the spread of infections in the cramped and unhygienic conditions where livestock is being raised (Ventola, 2015). The respective figure for the worldwide use of antimicrobials (mainly antibiotics) is 73% (Van Boeckel et al., 2019) Although direct use of antimicrobials for growth promotion has been banned recently in most countries, feed- and drinking water-additives containing antibiotics at subtherapeutic doses continue to be the norm in order to prevent the outbreak and spread of diseases. Projections show a further 67% expected global increase in the use of antibiotics in animal husbandry (Van Boeckel et al., 2015).

When people consume meat, dairy and egg products, they also ingest the antibiotics that had been given to the livestock as feed supplements. Antimicrobial resistance is now observed in humans and in farm animals (Tang et al.; 2017; Scott et al., 2018). Common drug-resistant pathogens include *Escherichia coli*, *Campylobacter* spp., nontyphoidal *Salmonella*spp. and *Staphylococcus aureus* (Van Boeckel et al., 2019). The World Health Organisation calls for limiting meat consumption as well as for reducing the use of antibiotics for food-producing animals (WHO, 2017b) with obvious co-benefits from these two measures. In the European Union, UK and USA, many other bodies concerned about human health and food quality have issues similar calls and there is also legislation in place; however, controlling the actions of the livestock industry globally remains a challenge.

Particularly concerning is also the fact that up to 90% of the antibiotics given to livestock cannot be absorbed by the animals' digestive systems and are excreted in their urine and faeces (Bartlett et al., 2013). These antibiotic-containing excrements penetrate the natural environment, including soils, freshwater bodies, groundwater and surface runoffs, affecting wildlife, including terrestrial and aquatic animals and any vegetation. Some microorganisms, such as bacteria and fungi, develop resistance mechanisms to the antimicrobials present in the natural environment resulting in antibiotic-resistant genes. Thus, these antimicrobial-resistant microorganisms can freely multiply and find ways to infuse new surroundings as well as pass between species through mobile genetic elements (CDC, 2019). Furthermore, continued use of one antibiotic can lead to resistance to multiple structurally unrelated antimicrobials through zoonotic (transferring between animals and humans) and sapronotic (existent in the wider environment) pathogenic bacteria (Raphaely et al., 2017). The US Centre for Disease Control and Prevention categorises antibiotic-resistant bacteria and fungi according to the level of concern for the human body as urgent, serious and concerning, and some are put on a watch list. Its 2019 Report (CDC, 2019) lists the following as urgent public health threats: Carbapenem-resistant *Acinetobacter*, *Candida auris*, *Clostridioides difficile*, Carbapenem-resistant Enterobacteriaceae and drug-resistant *Neisseria gonorrhoeae*.

The anthropogenic-driven environmental contamination results in wide geographical spread and transmission of antimicrobial resistance. Consequently, the scale of the problem becomes enormous and adds another sombre perspective to the planetary emergency. The abusive use of antibiotics in food-producing animals is likely to not only cut short their miraculous power, but also creates serious environmental contamination affecting all other species and exposing everyone and every single living organism at risk. Serious threats and consequences are already felt by this generation but are likely to amplify in the future (Morehead & Scarborough, 2018). The COVID-19 pandemic and the extensive use of antibiotics, biocides and disinfectants to combat and control the coronavirus disease are likely to add to the burden of antimicrobial resistance (Monnet & Harbarth, 2020; Rezasoltani et al., 2020; Strathdee et al., 2020) and the complexity of the remediation actions. By comparison, reduction in the desire for and intake of animal-based foods seems like a much easier and achievable task.

Conclusion

One of the lines from the Declaration of Geneva states: "I will maintain the utmost respect for human life" (WMA, 2017, para. 2). Can doctors truly commit to respecting human life without consideration of the food humans eat? The analysis in this chapter shows that within the current planetary emergency, the issues of food and human health are deeply intertwined. It is impossible to respect human life without valuing the biological and other ecological systems that produce food – the basic requirement for maintaining human life. With the Hippocratic Oath containing remarkable elements that are still relevant for medical ethics today (Askitopoulou & Vgontzas, 2018; Wiesing, 2020), it is time to bring food back into focus for the medical profession. The well-being of the human species can no longer be seen separately from the health of the planet and there is a role for all professions to play in this.

In this chapter, we discussed the link between food choices and diseases by making the case that a reduction in the intake of animal-based products has numerous benefits. Reducing the potential sources for antimicrobial resistance – the neglected looming global crisis exacerbated by COVID-19 (Getahun et al., 2020), is one of them. Preventing new zoonotic and non-communicable diseases is another important health front which can be managed with less addiction to meat, eggs, dairy and other animal-based products. Food has significantly contributed to the current planetary emergency but can also be a much-needed solution. Mainstreaming flexitarian diets can generate numerous co-benefits and reduce the burden of disease for the human race and the planet.

References

Aburto, N. J., Hanson, S., Gutierrez, H., Hooper, L., Elliott, P., & Cappuccio, F. P. (2013a). Effect of increased potassium intake on cardiovascular risk factors and disease: Systematic review and meta- analyses. *British Medical Journal, 346*, f1378. https://doi.org/10.1136/bmj.f1378

Aburto, N. J., Ziolkovska, A., Hooper, L., Elliott, P., Cappuccio, F. P., & Meerpohl, J. J. (2013b). Effect of lower sodium intake on health: Systematic review and meta-analyses. *British Medical Journal, 346*, f1326. https://doi.org/10.1136/bmj.f132

Aleksandrowicz, L., Green, R., Joy, E. J. M., Smith, P., & Haines, A. (2016). The impacts of dietary change on greenhouse gas emissions, land use, water use, and health: A systematic review. *Public Library of Science (PLoS) One, 11*(11), e0165797. https://doi.org/10.1371/journal.pone.0165797

American Heart Association. (2021). *Heart disease #1 cause of death rank likely to be impacted by COVID-19 for years to come.* https://www.eurekalert.org/pub_releases/2021-01/aha-hd012521.php

Askitopoulou, H., & Vgontzas, A. N. (2018). The relevance of the Hippocratic Oath to the ethical and moral values of contemporary medicine. Part I: The Hippocratic Oath from antiquity to modern times. *European Spine Journal, 27*(7), 1481–1490. https://doi.org/10.1007/s00586-017-5348-4

Australian Institute of Health and Welfare (AIHW). (2020). *Overweight and obesity.* https://www.aihw.gov.au/reports/australias-health/overweight-and-obesity

Australian Medical Association (AMA). (2006). *AMA adopts WMA Declaration of Geneva.* https://ama.com.au/media/ama-adopts-wma-declaration-geneva

Barclay, A. W., Petocz, P., McMillan-Price, J., Flood, V. M., Prvan, T., Mitchell, P., & Brand-Miller, J. C. (2008). Glycemic index, glycemic load, and chronic disease risk: A meta-analysis of observational studies. *The American Journal of Clinical Nutrition, 87*(3), 627–637. https://doi.org/10.1093/ajcn/87.3.627

Bartlett, J. G., Gilbert, D. N., & Spellberg, B. (2013). Seven ways to preserve the miracle of antibiotics. *Clinical Infectious Diseases, 56*(10), 1445–1550. https://doi.org/10.1093/cid/cit070

Bhupathiraju, S. N., Tobias, D. K., Malik, V. S., Pan, A., Hruby, A., Manson, J. E., … Hu, F. B. (2014). Glycemic index, glycemic load, and risk of type 2 diabetes: Results from 3 large US cohorts and an updated meta-analysis. *The American Journal of Clinical Nutrition, 100*(1), 218–232. https://doi.org/10.3945/ajcn.113.079533

Blanchflower, D. G., Oswald, A. J., & Stewart-Brown, S. (2012). *Is psychological well-being linked to the consumption of fruit and vegetables?* National Bureau of Economic Research (NBER). Working Paper, No. 18469. http://www.nber.org/papers/w18469.

Boehm, J. K., Williams, D. R., Rimm, E. B., Ryff, C., & Kubzansky, L. D. (2013). Association between optimism and serum antioxidants in the midlife in the United States study. *Psychosomatic Medicine, 75*(1), 2–10. https://doi.org/10.1097/PSY.0b013e31827c08a9

Bogueva, D., Marinova, D., & Raphaely, T. (2021). Influencing sustainable food-related behaviour changes: A case study in Sydney, Australia. In J. Bhattacharyya, M. K. Dash, C. Hewege, & M. S. Balaji (Eds.), *Social and sustainability marketing: A casebook for reaching your socially responsible consumers through marketing science.* Routledge.

Brazier, Y. (2020). *What is the Mediterranean diet?* Medical News Today. https://www.medical-newstoday.com/articles/149090

Burki, T. (2020). The origin of SARS-CoV-2. *The Lancet Infectious Diseases, 20*(9), 1018–1019. https://doi.org/10.1016/S1473-3099(20)30641-1

Centre for Disease Control and Prevention (CDC). (2017). *Zoonotic diseases.* https://www.cdc.gov/onehealth/basics/zoonotic-diseases.html

Centre for Disease Control and Prevention (CDC). (2019). *Antibiotic resistance threats in the United States 2019.* https://www.cdc.gov/drugresistance/pdf/threats-report/2019-ar-threats-report-508.pdf

Chu, Y., Yang, J., Shi, J., Zhang, P., & Wang, X. (2020). Obesity is associated with increased severity of disease in COVID-19 pneumonia: A systematic review and meta-analysis. *European Journal of Medical Research, 25*(1), 64. https://doi.org/10.1186/s40001-020-00464-9

Clark, M., & Tilman, D. (2017). Comparative analysis of environmental impacts of agricultural production systems, agricultural input efficiency, and food choice. *Environmental Research Letters, 12*(6), 064016. https://doi.org/10.1088/1748-9326/aa6cd5

Contreras García, E., & Zaragoza-Martí, A. (2020). Influence of food or food groups intake on the occurrence and/or protection of different types of cancer: Systematic review. *Nutrición Hospitalaria, 37*(1), 169–192. https://doi.org/10.20960/nh.02588

Dahl, W. J., Agro, N. C., Eliasson, A. M., Mialki, K. L., Olivera, J. D., Rusch, C. T., & Young, C. N. (2017). Health benefits of fiber fermentation. *Journal of the American College of Nutrition, 36*(2), 127–136. https://doi.org/10.1080/07315724.2016.1188737

Degeling, C., Lederman, Z., & Rock, M. (2016). Culling and the common good: Re-evaluating harms and benefits under the One Health paradigm. *Public Health Ethics, 9*(3), 244–254. https://doi.org/10.1093/phe/phw019

Editorial. (2013). The antibiotic alarm. *Nature, 495*(7440), 141. https://doi.org/10.1038/495141a

Ekmekioglu, C., Wallner, P., Kundi, M., Weisz, U., Haas, W., & Hutter, H.-P. (2018). Red meat, diseases, and healthy alternatives: A critical review. *Critical Reviews in Food Science and Nutrition, 58*(2), 247–261. https://doi.org/10.1080/10408398.2016.1158148

Ellwanger, J. H., Veiga, A., Kaminski, V. L., Valverde-Villegas, J. M., Freitas, A., & Chies, J. (2021). Control and prevention of infectious diseases from a One Health perspective. *Genetics and Molecular Biology, 44*(Suppl 1), e20200256. https://doi.org/10.1590/1678-4685-GMB-2020-0256

FAOSTAT. (2021). *Live animals*. Food and Agriculture Organisation of the United Nations. http://www.fao.org/faostat/en/#data/QA

Food and Agriculture Organisation of the United Nations (FAO), International Fund for Agricultural Development (IFAD), United Nations International Children's Emergency Fund (UNICEF), World Food Programme (WFP) & World Health Organisation (WHO). (2019). *The State of Food Security and Nutrition in the World 2019. Safeguarding against economic slowdowns and downturns*. http://www.fao.org/3/ca5162en/ca5162en.pdf

Getahun, H., Smith, I., Trivedi, K., Paulin, S., & Balkhy, H. H. (2020). Tackling antimicrobial resistance in the COVID-19 pandemic. *Bulletin of the World Health Organization, 98*(7), 442–442A. https://doi.org/10.2471/BLT.20.268573

Gibb, R., Redding, D. W., Chin, K. Q., Donnelly, C. A., Blackburn, T. M., Newbold, T., & Jones, K. E. (2020). Zoonotic host diversity increases in human-dominated ecosystems. *Nature, 584*, 398–402. https://doi.org/10.1038/s41586-020-2562-8

Global Panel on Agriculture and Food Systems for Nutrition. (2016). *Food systems and diets: Facing the challenges of the 21st century*. London, UK. https://www.glopan.org/foresight1/

Greenwood, D. C., Threapleton, D. E., Evans, C. E., Cleghorn, C. L., Nykjaer, C., Woodhead, C., & Burley, V. J. (2013). Glycemic index, glycemic load, carbohydrates, and type 2 diabetes: Systematic review and dose-response meta-analysis of prospective studies. *Diabetes Care, 36*(12), 166–171. https://doi.org/10.2337/dc13-0325

Haddad, E. H. (2018). Vegetarian diet and risk of cardiovascular disease. In W. J. Craig (Ed.), *Vegetarian nutrition and wellness* (pp. 45–70). CRC Press.

Haider, N., Rothman-Ostrow, P., Osman, A. Y., Arruda, L. B., Macfarlane-Berry, L., Elton, L., … Kock, R. A. (2020). COVID-19-zoonosis or emerging infectious disease? *Frontiers in Public Health, 8*, 596944. https://doi.org/10.3389/fpubh.2020.596944

Haydon, D. T., Cleaveland, S., Taylor, L. H., & Laurenson, M. K. (2002). Identifying reservoirs of infection: A conceptual and practical challenge. *Emerging Infectious Diseases, 8*(12), 1468–1473. https://doi.org/10.3201/eid0812.010317

Hinchcliffe, S. (2015). More than one world, more than one health: Re-configuring interspecies health. *Social Science and Medicine, 129*, 28–35. https://doi.org/10.1016/j.socscimed.2014.07.007

Hu, D., Huang, J., Wang, Y., Zhang, D., & Qu, Y. (2014). Fruits and vegetables consumption and risk of stroke: A meta-analysis of prospective cohort studies. *Stroke, 45*(6), 1613–1619. https://doi.org/10.1161/STROKEAHA.114.004836

Indian Council of Medical research (ICMR), & National Institute of Nutrition (NIN). (2019). *Share of fruits and vegetables in tackling CVDs and NCDs (especially diabetes, heart attack, stroke and cancer) in Indian context*. Hyderabad. https://www.nin.res.in/brief/Fruits_and_Vegetables.pdf

Institute of Medicine. (2003). *Microbial threats to health: Emergence, detection, and response*. The National Academies Press. https://doi.org/10.17226/10636

Kye, S. Y., & Park, K. (2014). Health-related determinants of happiness in Korean adults. *International Journal of Public Health, 59*, 731–738. https://doi.org/10.1007/s00038-014-0588-0

Loef, M., & Walach, H. (2012). Fruit, vegetables and prevention of cognitive decline or dementia: A systematic review of cohort studies. *The Journal of Nutrition, Health & Aging, 16*(7), 626–630. https://doi.org/10.1007/s12603-012-0097-x

Mackenzie, J. S., & Smith, D. W. (2020). COVID-19: A novel zoonotic disease caused by a coronavirus from China: What we know and what we don't. *Microbiology Australia, MA20013.* https://doi.org/10.1071/MA20013

Malik, V. S., Li, Y., Tobias, D. K., Pan, A., & Hu, F. B. (2016). Dietary protein intake and risk of type 2 diabetes in US men and women. *American Journal of Epidemiology, 183*(8), 715–728. https://doi.org/10.1093/aje/kwv268

Marsh, K., Zeuschner, C., & Saunders, A. (2012). Health implications of a vegetarian diet: A review. *American Journal of Lifestyle Medicine, 6*(3), 250–267. https://doi.org/10.1177/1559827611425762

Marsh, K., Zeuschner, C., & Saunders, A. (2016). Red meat and health: Evidence regarding red meat, health, and chronic disease risk. In T. Raphaely & D. Marinova (Eds.), *Impact of meat consumption on health and environmental sustainability* (pp. 131–177). IGI Global.

Marshall, P., & Marinova, D. (2019). Health benefits of eating more plant foods and less meat. In D. Bogueva, D. Marinova, T. Raphaely, & K. Schmidinger (Eds.), *Environmental, health and business opportunities in the new meat alternatives market* (pp. 38–61). IGI Global.

Mason-D'Croz, D., Bogard, J. R., Sulser, T. B., Cenacchi, N., Dunston, S., Herrero, M., & Wiebe, K. (2019). Gaps between fruit and vegetable production, demand, and recommended consumption at global and national levels: An integrated modelling study. *The Lancet Planetary Health, 3*(7), e318–e329. https://doi.org/10.1016/S2542-5196(19)30095-6

McFarlane, R. A., Sleigh, A. C., & McMichael, A. J. (2013). Land-use change and emerging infectious disease on an island continent. *International Journal of Environmental Research and Public Health, 10*(7), 2699–2719. https://doi.org/10.3390/ijerph10072699

Medawar, E., Huhn, S., Villringer, A., & Witte, A. V. (2019). The effects of plant-based diets on the body and the brain: A systematic review. *Translational Psychiatry, 9,* 226. https://doi.org/10.1038/s41398-019-0552-0

Michael, C. A., Dominey-Howes, D., & Labbate, M. (2014). The antimicrobial resistance crisis: Causes, consequences, and management. *Frontiers in Public Health, 2,* 145. https://doi.org/10.3389/fpubh.2014.00145

Michalon, J. (2020). Accounting for One Health: Insights from the social sciences. *Parasite, 27,* 56. https://doi.org/10.1051/parasite/2020056

Middleton, J., Reintjes, R., & Lopes, H. (2020). Meat plants – A new front line in the covid-19 pandemic: These businesses failed in their duty to workers and the wider public health. *BMJ, 370,* m2716. https://doi.org/10.1136/bmj.m2716

Monnet, D. L., & Harbarth, S. (2020). Will coronavirus disease (COVID-19) have an impact on antimicrobial resistance? *Euro Surveillance, 25*(45), 2001886. https://doi.org/10.2807/1560-7917.ES.2020.25.45.2001886

Morehead, M. S., & Scarborough, C. (2018). Emergence of global antibiotic resistance. *Primary Care, 45*(3), 467–484. https://doi.org/10.1016/j.pop.2018.05.006

Mukpo, A. (2020). *As wild areas become farmland, species that carry diseases flourish.* Mongabay. https://news.mongabay.com/2020/08/as-wild-areas-become-farmland-species-that-carry-diseases-flourish/

Nadeau, K. J., Anderson, B. J., Berg, E. G., Chiang, J. L., Chou, H., Copeland, K. C., … Zeitler, P. (2016). Youth-onset type 2 diabetes consensus report: Current status, challenges, and priorities. *Diabetes Care, 39*(9), 1635–1642. https://doi.org/10.2337/dc16-1066

National Research Council (US). (2009). *Committee on achieving sustainable global capacity for surveillance and response to emerging diseases of zoonotic origin.* National Academies Press.

North, M. (2002). *Greek medicine: "I swear by Apollo Physician…": Greek medicine from the Gods to Galen.* US National Library of Medicine. http://www.nlm.nih.gov/hmd/greek/greek_oath.html

Otaegui-Arrazola, A., Amiano, P., Elbusto, A., Urdaneta, E., & Martínez-Lage, P. (2014). Diet, cognition, and Alzheimer's disease: Food for thought. *European Journal of Nutrition, 53*(1), 1–23. https://doi.org/10.1007/s00394-013-0561-3

Our World in Data. (n.d.). *Number of animals slaughtered for meat, World, 1961 to 2018.* https://ourworldindata.org/grapher/animals-slaughtered-for-meat

Petersson, S., & Philippou, E. (2016). Mediterranean diet, cognitive function, and dementia: A systematic review of the evidence. *Advances in Nutrition, 7*(5), 889–904. https://doi.org/10.3945/an.116.012138

Phillips, C. (2020). *Coronavirus: Live animals are stressed in wet markets, and stressed animals are more likely to carry diseases.* The Conversation. https://theconversation.com/coronavirus-live-animals-are-stressed-in-wet-markets-and-stressed-animals-are-more-likely-to-carry-diseases-135479

Piqueras, J. A., Kuhne, W., Vera-Villarroel, P., van Straten, A., & Cuijpers, P. (2011). Happiness and health behaviours in Chilean college students: A cross-sectional survey. *BMC Public Health, 7*(11), 443. https://doi.org/10.1186/1471-2458-11-443

Poon, L. L., Guan, Y., Nicholls, J. M., Yuen, K. Y., & Peiris, J. S. (2004). The aetiology, origins, and diagnosis of severe acute respiratory syndrome. *The Lancet Infectious Diseases, 4*(11), 663–671. https://doi.org/10.1016/S1473-3099(04)01172-7s

Poore, J., & Nemecek, T. (2018). Reducing food's environmental impacts through producers and consumers. *Science, 360,* 987–992. https://doi.org/10.1126/science.aaq0216

Population Reference Bureau. (2018). *2018 World population data sheet.* https://www.prb.org/wp-content/uploads/2018/08/2018_WPDS.pdf

Rajaram, S., Jones, J., & Lee, G. J. (2019). Plant-based dietary patterns, plant foods, and age-related cognitive decline. *Advances in Nutrition, 10*(Suppl 4), S422–S436. https://doi.org/10.1093/advances/nmz081

Ramasamy, R. (2014). Zoonotic malaria – Global overview and research and policy needs. *Frontiers in Public Health, 2,* 123. https://doi.org/10.3389/fpubh.2014.00123

Raphaely, T., Marinova, D., & Marinova, M. (2017). The future of antibiotics and meat. In M. Khosrow-Pour (Ed.), *Public health and welfare: Concepts, methodologies, tools, and applications* (pp. 1335–1357). IGI Global.

Rezasoltani, S., Abbas, Y., Behzad, H., Hamid, A. A., & Reza, M. (2020). Antimicrobial resistance as a hidden menace lurking behind the COVID-19 outbreak: The global impacts of too much hygiene on AMR. *Frontiers in Microbiology, 11.* https://doi.org/10.3389/fmicb.2020.590683

Ritchie, H., & Roser, M. (2019). *Land use.* https://ourworldindata.org/land-use

Ritchie, H., & Roser, M. (2020). *Environmental impacts of food production.* https://ourworldindata.org/environmental-impacts-of-food

Ross, S. M. (2018). The nutrition-brain connection: Nutritional status and cognitive decline. *Holistic Nursing Practice, 32*(3), 169–171. https://doi.org/10.1097/HNP.0000000000000270

Satija, A., Bhupathiraju, S. N., Donna Spiegelman, D., Chiuve, S. E., Manson, J. E., Willett, W., … Hu, F. B. (2017). Healthful and unhealthful plant-based diets and the risk of coronary heart disease in U.S. adults. *Journal of the American College of Cardiology, 70*(4), 411–422. https://doi.org/10.1016/j.jacc.2017.05.047

Scott, A. M., Beller, E., Glasziou, P., Clark, J., Ranakusuma, R. W., Byambasuren, O., … Del Mar, C. (2018). Is antimicrobial administration to food animals a direct threat to human health? A rapid systematic review. *International Journal of Antimicrobial Agents, 52*(3), 316–323. https://doi.org/10.1016/j.ijantimicag.2018.04.005

Sengupta, S., Chattopadhyay, M. K., & Grossart, H. P. (2013). The multifaceted roles of antibiotics and antibiotic resistance in nature. *Frontiers in Microbiology, 4,* 47. https://doi.org/10.3389/fmicb.2013.00047

Shang, X., Scott, D., Hodge, A. M., English, D. R., Giles, G. G., Ebeling, P. R., & Sanders, K. M. (2016). Dietary protein intake and risk of type 2 diabetes: Results from the Melbourne Collaborative Cohort Study and a meta-analysis of prospective studies. *The American Journal of Clinical Nutrition, 104*(5), 1352–1365. https://doi.org/10.3945/ajcn.116.140954

Shepon, A., Eshel, G., Noor, E., & Milo, R. (2018). The opportunity cost of animal based diets exceeds all food losses. *Proceedings of the National Academy of Sciences of the United States of America (PNAS), 15*(15), 3804–3809. https://doi.org/10.1073/pnas.1713820115

Spinney, L. (2020). *Wuhan virologist says more bat coronaviruses capable of crossing over*. The Guardian. https://www.theguardian.com/world/2020/dec/04/wuhan-virologist-warns-more-bat-coronaviruses-capable-of-crossing-over-covid-beyond-china

Springmann, M., Godfray, H. C. J., Rayner, M., & Scarborough, P. (2016). Analysis and valuation of the health and climate change cobenefits of dietary change. *Proceedings of the National Academy of Sciences of the United States of America (PNAS), 113*(15), 4146–4151. https://doi.org/10.1073/pnas.1523119113

Strathdee, S. A., Davies, S. C., & Marcelin, J. R. (2020). Confronting antimicrobial resistance beyond the COVID-19 pandemic and the 2020 US election. *The Lancet, 396*(10257), 1050–1053. https://doi.org/10.1016/S0140-6736(20)32063-8

Stuchin, M., Machalaba, C. C., & Karesh, W. B. (2016). Vector-borne diseases: Animals and patterns. In *Global health impacts of vector-borne diseases: Workshop summary* (pp. 167–181). National Academies Press.

Tan, M., He, F. J., & MacGregor, G. A. (2020). Obesity and covid-19: The role of the food industry. *BMJ, 369*, m2237. https://doi.org/10.1136/bmj.m2237

Tang, K. L., Caffrey, N. P., Nóbrega, D. B., Cork, S. C., Ronksley, P. E., Barkema, H. W., … Ghali, W. A. (2017). Restricting the use of antibiotics in food-producing animals and its associations with antibiotic resistance in food-producing animals and human beings: A systematic review and meta-analysis. *Lancet Planet Health, 1*(8), e316–e327. https://doi.org/10.1016/S2542-5196(17)30141-9

Te Beest, D. E., Hagenaars, T. J., Stegeman, J. A., Koopmans, M. P. G., & van Boven, M. (2011). Risk based culling for highly infectious diseases of livestock. *Veterinary Research, 42*, 81. https://doi.org/10.1186/1297-9716-42-81

Tuso, P. J., Ismail, M. H., Ha, B. P., & Bartolotto, C. (2013). Nutritional update for physicians: Plant-based diets. *The Permanente Journal, 17*(2), 61–66. https://doi.org/10.7812/TPP/12-085

Van Boeckel, T. P., Brower, C., Gilbert, M., Grenfell, B. T., Levin, S. A., Robinson, T. P., Teillant, A., & Laxminarayan, R. (2015). Global trends in antimicrobial use in food animals. *Proceedings of the National Academy of Sciences of the United States of America (PNAS), 112*(18), 5649–5654. https://doi.org/10.1073/pnas.1503141112

Van Boeckel, T. P., Pires, J., Reshma, S., Zhao, C., Song, J., Criscuolo, N. G., … Laxminarayan, R. (2019). Global trends in antimicrobial resistance in animals in low- and middle-income countries. *Science, 365*(6459), eaaw1944. https://doi.org/10.1126/science.aaw1944

Ventola, C. L. (2015). The antibiotic resistance crisis. Part 1: Causes and threats. *Pharmacy and Therapeutics, 40*(4), 277–283.

Wei, H., Gao, Z., Liang, R., Li, Z., Hao, H., & Liu, X. (2016). Whole-grain consumption and the risk of all-cause, CVD and cancer mortality: A meta-analysis of prospective cohort studies. *British Journal of Nutrition, 116*(03), 514–525. https://doi.org/10.1017/S0007114516001975

Wiesing, U. (2020). The Hippocratic Oath and the Declaration of Geneva: Legitimisation attempts of professional conduct. *Medicine, Health Care, and Philosophy, 23*(1), 81–86.

White, B. A., Horwath, C. C., & Conner, T. S. (2013). Many apples a day keep the blues away – daily experiences of negative and positive affect and food consumption in young adults. *British Journal of Health Psychology, 18*(4), 782–798. https://doi.org/10.1111/bjhp.12021

Willett, W., Rockström, J., Loken, B., Springmann, M., Lang, T., Vermeulen, S., … Murray, C. J. L. (2019). Food in the Anthropocene: The EAT–Lancet Commission on healthy diets from sustainable food systems. *The Lancet, 393*(10170), 447–492. https://doi.org/10.1016/S0140-6736(18)31788-4

World Cancer Research Fund/American Institute for Cancer Research (WCRF/AICR). (2018). *Diet, nutrition, physical activity and cancer: A global perspective*. Continuous Update Project Expert Report 2018 https://www.wcrf.org/dietandcancer/contents

World Health Organisation (WHO). (2015). *Q&A on the carcinogenicity of the consumption of red meat and processed meat*. http://www.who.int/features/qa/cancer-red-meat/en/

World Health Organisation (WHO). (2017a). *WHO guidelines on use of medically important antimicrobials in food-producing animals.* https://www.who.int/foodsafety/areas_work/ antimicrobial-resistance/cia_guidelines/en/

World Health Organisation (WHO). (2017b). *Cardiovascular diseases (CDVs): Key facts.* http:// www.who.int/en/news-room/fact-sheets/detail/cardiovascular-diseases-(cvds)

World Health Organisation (WHO). (2018). *Noncommunicable diseases.* https://www.who.int/ news-room/fact-sheets/detail/noncommunicable-diseases

World Health Organisation (WHO). (2019). *Increasing fruit and vegetable consumption to reduce the risk of noncommunicable diseases.* e-Library of Evidence for Nutrition Actions (eLENA). https://www.who.int/elena/titles/fruit_vegetables_ncds/en/

World Health Organisation (WHO). (2020). *Zoonoses.* https://www.who.int/news-room/ fact-sheets/detail/zoonoses

World Health Organisation (WHO). (2021a). *Neglected zoonotic diseases.* https://www.who.int/ neglected_diseases/diseases/zoonoses/en/

World Health Organisation (WHO). (2021b). *International Year of Fruits and Vegetables 2021.* http://www.fao.org/fruits-vegetables-2021/en/

World Health Organisation (WHO), Food and Agriculture Organisation of the United Nations (FAO), & World Organisation for Animal Health. (2004). *Report of the WHO/FAO/OIE joint consultation on emerging zoonotic diseases.* https://www.oie.int/doc/ged/D5681.PDF

World Medical Association (WMA). (2017). *WMA Declaration of Geneva.* https://www.wma.net/ policies-post/wma-declaration-of-geneva/

Yang, J., Tian, C., Chen, Y., Zhu, C., Chi, H., & Li, J. (2020). Obesity aggravates COVID-19: An updated systematic review and meta-analysis. *Journal of Medical Virology, 17.* https://doi. org/10.1002/jmv.26677

Zhao, X., Gang, X., He, G., Li, Z., Lv, Y., Han, Q., & Wang, G. (2020). Obesity increases the severity and mortality of influenza and COVID-19: A systematic review and meta-analysis. *Frontiers in Endocrinology, 11,* 595109. https://doi.org/10.3389/fendo.2020.595109

Chapter 11
Generation Z and Food Choices

Abstract Generation Z, the largest population cohort on this planet, is already proving to be environmentally aware and socially active as demonstrated with its support for the climate strikes and other related movements. The chapter analyses how these young people engage with food and diets – issues that are gradually becoming a political, environmental, social, cultural and public health agenda. It describes the food choices of Generation Z posing the question whether there is a transitioning to a better planetary diet. Using empirical evidence from Australia the chapter describes some of the trends and lessons given by these young people concluding that they hold key answers about the planet's future.

Food and diets have become a political, environmental, social, cultural and public health issue, but how are they perceived by Generation Z (Gen Z)? This generation is already proving to be highly environmentally aware and socially active as demonstrated with its support for the climate strikes and social movements. This chapter analyses the food choices of Gen Z posing the question whether it is transitioning to a better planetary diet. Gen Z holds the hopes for the future on this planet but will also have to deal with the consequences from human-induced climate change and environmental deterioration. Are these young people ready? How do they see their role in a food transition? We refer to empirical evidence from Australia to describe some of the trends and lessons given by these young people.

Gen Z and the Planetary Emergency

History is dependent on the new generation to write a new chapter. LaMelo Ball (American professional basketball player, Gen Z, born 2001)

The air was filled with an expressly nervous but uplifting spirit and a clear enthusiasm gushing from the young people around us. We were catching the train to the heart of Sydney thrilled to be able to join the marches of Gen Z – the evolving

© The Author(s), under exclusive license to Springer Nature Singapore Pte Ltd. 2022
D. Marinova, D. Bogueva, *Food in a Planetary Emergency*, https://doi.org/10.1007/978-981-16-7707-6_11

powerful young cohort, and observe close up their desire to change the world for the better. It was 2019, in the middle of the "School Strike 4 Climate". We reached the City Town Hall and found ourselves in the centre of the climate strike movement. The train station was suddenly packed with young climate activists carrying a sea of placards, burning with passion, emotions, enthusiasm and a degree of anger, but cheerful, motivated and optimistic self-born environmentalists. They were bright-eyed and jaunty young people who were already taking the planet's future into their own hands. Gen Z was ready to fight, ready to achieve a change. You could feel this everywhere in the air. They were all ready to make the impossible possible, to be heard, to influence political action and to make a real difference now and for their future lives, and for the lives of their children. Gen Z was there with a clear vision – it wanted to give an ultimatum to find solutions to save the world. The ultimatum was addressed to the world's political leaders, who were running away from their climate change and environmental responsibilities or pretended not to see the problems.

Greta Thunberg, the then 16-year-old Swedish school strike activist, was not in Sydney but her words from the Davos World Economic Forum resonated in the air: "Our house is on fire... I don't want your hope. I want you to panic...and act...." (Workman, 2019, para. 1). These strong words from a small girl intended to slam the global leaders' inaction on climate change and urge them to fear the crisis and act. Greta's words reverberated in our ears piercing our consciousness like a dagger.

At the other end of Australia, in the Western Australian coastal city of Bunbury, the then 10-year schoolgirl Bella Burgemeister (2019) had already written a book challenging her schoolmates and other students to implement simple changes that help achieve the United Nations Sustainable Development Goals (SDGs). She wrote: "There is something wrong with the way we are treating our planet. There is something wrong with the way we are treating each other. I am not the only one who thinks we have a problem" (Burgemeister, 2019, p. 2). Bella then went on present-ing her sustainability challenges to over 6000 students around Australia, with all proceeds from the sales of her book going into sustainable development projects (Valuing Children Initiative, 2021).

In May 2021, thousands of students gathered again in Sydney's CBD to resume student climate strikes after a year of devastation caused by the COVID-19 pandemic. The climate actions in Australia and around the world did not stop during this time. Gen Z found new forms of expression and acting (Storch et al., 2021). It continued to be unstoppable. The voices of youth did not disappear during the pandemic but moved to their digital native space online creating an even stronger and powerful global com-munity united in "clicktivism" (Finnegan, 2020). They were mobilising online at #ClimateStrikeOnline, sharing memes and hashtags, participating in Fridays for Future (2020) digital strikes, raising their voice on the Internet with specially designed toolkits (UNICEF, n.d.), driving the United Nations Youth Environment Assembly's talks (UNEP, 2021), live-streaming, hosting webinars (IIED, 2020) and creating Climate Bots (U report, n.d.). These were some of Gen Z's new tactics.

For governments and politicians, the actions on improving planetary health (Marinova & Bogueva, 2019) are a bit of a flinch, especially when economic and political targets are more visible than the "imaginary" future climate crisis. Politicians rarely think that their actions represent a betrayal of the young people's future. "How dare you – you have stolen my dreams and my childhood… All you can talk about is money and fairy tales of eternal economic growth" (NPR Staff, 2019, para. 4). Whoever Greta Thunberg blasted at the United Nations Climate Action Summit in September 2019 should trigger real panic and response from them. We did not see such a response, at least not in Australia and the rest of the developed world. There was not any meaningful political action, with the world using the COVID-19 pandemic as a way to deflect attention. The right and clear messages of care about the climate were not sent to the young people. Gen Z read such a response as denying the fact that humanity is facing the climate change devastations. With inaction, we will not be able to win the climate change race (UN, 2020) and Gen Z sees this clearly. Worse, by supporting industries, such as fossil-fuel based energy, livestock and clearing of native vegetation, the Australian government was making things uneasier for the future.

The meek COVID-19 situation in Australia in May 2021 made it possible for the school strikes to resume. Students rallied and voiced their anger across 50 Australian cities and towns with thousands protesting against climate inaction, angry with the government's investments in gas and coal and calling for transition to an economy away from fossil fuels (Young, 2021). However, Gen Z's actions did not stop here. They took the Australian Government to the Federal Court about a decision to approve the extension of a coal mine arguing that this would exacerbate climate change and cause serious future harm to them (Slezak & Timms, 2021). Gen Z won! Bella Burgemeister was among this group of eight Australian teenagers who won the landmark case against Australia's environment minister on behalf of all young Australians. The Court ruled out that the Australian Government had a legal duty to protect and not cause harm to the young people of Australia by exacerbating climate change when approving coal mining projects – a ruling seen by legal experts as opening up a "big crack in the wall" for future climate change-based litigation (Slezak & Timms, 2021).

Young people worldwide are environmentally aware and motivated by climate change. Those who think about the current situation are utterly traumatised by it and particularly by inheriting the burden left from the previous generations, with which they will need to deal. They are distraught about our planet's future, they require urgency and immediately addressing of the causes in order to halt the ecological destruction. Gen Z is not in despair or depressed because of what we have done, but rather irritated and angry by the lack of action and the resulting prospects for them. They feel that their house is truly burning and they are eager finding solutions to put the fire down.

People like Greta Thunberg, Bella Burgemeister and many others began spontaneously their climate crusades for the future of the planet. Their voices united strong protest networks across the globe (see Fig. 11.1) and inspired many young people to concentrate on actions about climate in their hearts and brains. They took the environmental agenda in their hands but do they also understand the challenges about food production? This is what we investigate in the chapter starting first by describing the characteristics of Gen Z, then moving into their food behaviour and attitudes towards meat and alternative proteins. Is Gen Z likely to influence broad segments

Fig. 11.1 Fridays for Future strikes poster for Sofia, Bulgaria. (Source: Authors)

of society, their parents and grandparents, people regardless of age, walks of life or political affiliations, to deal with food in a planetary emergency?

Are You Ready, Gen Z?

It is unfortunately true that our generation and that of your parents have left you with a big mess that will now be yours to clean up: wars, budget challenges, pollution, global warming, battles of health care, natural disasters. They're all there for you. We're willing those to you. Are you ready? John Morgridge (ex-CEO and chairman of the board of Cisco Systems, born 1933)

It is true that we cannot expect the world to get better by itself. The climate emergency we are facing happens to be one of the many challenges Gen Z needs to handle and they are surely and visibly embracing the imperative to act before it is too late. Nothing is more singular about Gen Z than their determination to find solutions to make things better.

Comprising 30% of the world's population, Gen Z is a realistic generation living in an incredibly unique time in history. Born between 1995 and 2010, Gen Z represents nearly 2 billion of the global population. It is the most culturally diverse generation in countries, such as USA and Australia (Deloitte, 2019), and recently started to attract increasing attention (Seemiller & Grace, 2019). A large fraction of Gen Z is concentrated in low- and middle-income countries located particularly in rural areas in Africa, South and East Asia (IFAD, 2019).

Shaped differently than any other previous generations, Gen Z is the most connected ever, and unique in its intense, intuitive engagement with technology. Living in a different world, with no limits to being digitally globally connected, communicating and sharing ideas with anyone at any time with each other (The Hartman Group, 2018), there are no secrets for them if something happens in the world in their vicinity or 1000+ km away. Gen Z will find out the details through social media or friends, particularly if it immediately affects them. Using energy and dynamism, their engagement can yield boundless results. They can do and be anything – employees, workers, farmers, food systems transformers, entrepreneurs of tomorrow and the drivers for inequality mitigation, economic development, employment opportunities, poverty reduction, food and nutrition security, conflict resolutions, fighters for change and global challenge activists (Ayele et al., 2017; IFAD, 2019; Yami et al., 2019; Glover & Sumberg, 2020). They are experiencing and being shaped by a rate and types of change dramatically different from before (IFAD, 2019).

This generation started their path and continues its journey in turbulence and instability, including the frightening and unsafe world of the 9/11 war on terrorism, conflicts and wars in Africa – Libya, Sierra Leone, Algeria, Somalia, Kongo, Yemen and Sudan, the Middle East –Syria and Iraq, Asia – Afghanistan, Chechnya, Hong Kong, Myanmar and the Philippines, and South America – Venezuela and Mexico. They grew up in times of socio-economic stagnation, when their parents lost their jobs and houses during the 2007–2009 global financial crisis (RBA, n.d.) or in times when their families were forcibly displaced, joined fleeing refugees in seeking asylum and a new home because of poverty, political unrest, gang violence, other serious circumstances or natural disasters, making many of them climate change migrants (Ida, 2021; Reid, 2021; UNHCR, 2021). Their list of experiences includes mass school shootings (Keneally, 2019), school violence and bulling (UNESCO, 2019), including cyber bullying, following them in their usual 24/7 habitual place in the social media and the Net, combined with higher rates of anxiety, depression and youth suicides (WHO, 2021).

Gen Z came of age in socially diverse family structures, including multi-racial, single-parent, rainbow and same-sex households, with blurred gender roles. These young people are less fazed than previous generations by differences in race, sexual orientation or religion showing unprecedented and even revolutionary understanding of the Lesbian, Gay, Bisexual, Transgender, Queer or Questioning and Intersex (LGBTQI) community, marriage equality and intolerance of racism (Deloitte, 2021). Such attitudes influence their actions, including Black Lives Matter (Borg &

Aleah, 2020), the Arab Spring and others aimed to change perceptions of people and the world we live in. A 2021 global study revealed that 40% of Gen Z has created social media content relating to an environmental, human rights, political or social issue and 30% participated in a public demonstration, some protest or march (Deloitte, 2021). Furthermore, 40% of Gen Z believes that COVID-19 will create a world where individuals are more likely to act on environmental issues (Deloitte, 2021).

Furthermore, Gen Z is prompted to take action with 26% reporting that they have contacted an elected representative to express views about a cause of concern to them (Deloitte, 2021). The young Pakistani activist Malala Yousafzai (see Fig. 11.2) was shot by the Talibans for promoting girls' education. She survived and continues to give hope for change and follows closely the developments of the Taliban forced

Fig.   11.2 Malala   Yousafzai   and   Greta   Thunberg.   (Credit: Malala's   Twitter account) **Malala** ✓

@Malala

She's the only friend I'd skip school for.

takeover of Afghanistan (SBS News, 2021). Only at the age of 17, she became the youngest person in history to win a Nobel prize in 2014. After surviving the Parkland, Florida school shooting, the gun control activist Emma Gonzalez became an advocate who co-founded the gun-control advocacy group #Never Again (Alter, 2018). The 2019 Time Person of the Year, the then 16-year-old climate change activist Greta Thunberg is another poignant example of social engagement leading Gen Z in climate activism (Alter et al., 2019).

Now when more than ever humanity is changing the face of the planet at an increasing rate, contributing to an unprecedented environmental breakdown (Laybourn-Langton et al., 2019), Gen Z is inheriting a scarier world shaped by climate change calamities and disasters happening daily around the globe. Certainly, these millions of Gen Z mobilised by the School Strikes for Climate and Fridays for Future campaigns are not better equipped to face the climate emergency challenges on behalf of the rest of the society but remain eager to fight and search for solutions. This Gen Z is not going to sit back and wait for others to develop a way out but will find means resonating with their beliefs, moral values and perceptions about what is right and what is wrong without missing opportunities to implement change. It is showing the power of youth and the upcoming change (see Fig. 11.3).

In fact, since the hippie movement of young people in the 1960s, there has not been another combined world movement behind a significant cause. In the 1960s it was to stop the war to save humankind; now it is to stop humankind to save the planet and its future. Gen Z's attitudes and expectations for transparency, accountability and trust, are poised to shape the next normal and they are ready to face what the previous generations are willing to them.

Fig. 11.3 A picture of Greta Thunberg waking up the world. (Credit: Getty (free images))

Solution-Driven Gen Z

Vision without action is merely a dream, action without vision is merely passing time, but action with vision can change the world. Nelson Mandela (South African anti-apartheid revolutionary, statesman and philanthropist, born 1918)

The realm of Gen Z is a world in which people can be hurt more easily, have mental health issues more frequently but are also more connected. They live in a world where being different is normal and often celebrated, discrimination is not tolerated. Gen Z wants to make a difference in the future and hopefully change the often scary and threatening predictions about the world into something beautiful and loving. Many of them are already fundraising for particular causes or running not-for-profit organisations. An example is Thomas King, a social entrepreneur and founder of Food Frontier, an alternative proteins company, whose ambition was to create a more sustainable, nutritious and future-proof food supply (Lethlean, 2020). Another example is Hunter Swisher and Erin Knabe, the co-founders of Phospholutions company, who turned an idea to enhance root growth in plants into a viable product developing a process that reduces fertiliser consumption and phosphorus runoff (Phospholutions, 2021). This solution reflects their belief that sustainable production and responsible use of phosphorus are fundamental to support a growing population (Phospholutions, 2021). Gen Z is craving opportunities for longer and deeper involvement and work toward creating a meaningful change. They already have their own ideas how they can make the world a better place. Instead of being followers, they prefer to start tackle the problems themselves. They cannot be engaged in a system that does not seem to work for them and focus on issues that matter, such as conserving water in, for examples, bathrooms (Seemiller & Grace, 2017).

Gen Z is not only disrupting and questioning the status quo, they are working around it to find ways to make a difference, including protesting in front of the parliament to fight for climate change, fighting for female education rights or responding to health challenges. They constantly work toward solutions and no complex problems are out of reach. For example, a group of Australian high school students managed to recreate the key ingredients of a life-saving HIV and malaria drug in their lab for a small fraction of the cost (Davey, 2016; Hunjan, 2016). During the COVID-19 lockdown in UK, Gen Z aged between 16 and 20 has been behind a boom in new businesses with a 72% increase in registrations as sole traders while all other age groups experienced a decline (Hughes, 2021). Many examples exist in the area of food. The Gen Z representatives George Peppou and Nick Noakesmith from the cultured meat start-up Vow saw the problem with livestock, and worked toward finding solutions creating an entirely new way of growing meat from animal cells, because for them making a difference comes with a business model and innovation. Instead of breeding and slaughtering animals, a finished meal can be produced in 6 weeks from a handful of animal cells (Vow, 2020). Me & the Bees Lemonade was established in Austin, Texas by Mikaila Ulmer at the age of 4 to save the honeybees (Overdeep, 2020). The breakfast overnight-oats brand Oatsu was founded by Lauren O'Donnell, initially as an Instagram-based company.

Influenced by the gut-health knowledge of the cultured grain amazake, passed to Lauren through her mother's Japanese roots, Oatsu was established in 2020 in UK as a business devoted to creating plant-based, gut-friendly breakfast options (Oatsu, 2021). Another UK-based Gen Z business is that of Ayesha Grover, the founder of strp'd. Her company is dedicated to boosting the immunity and improving gut health through producing and selling online nut-, gluten- and allergen-free tigernut flour – a starchy powder made from root vegetables, for baking (Enterprise Nation, 2021; strp'd, n.d.).

Each generation brings its own individual perceptions, judgments and actions on particular issues, but never before has the world faced the severity, urgency and complexity of the challenges we have today. The size of the global population is reinforcing the seriousness of the problems, especially when it comes to food. Gen Z is set to influence collective and political responses. Their dietary choices are influenced by their close social connections and conform with the behaviour of others but also exhibit generation-specific shared cultural expectations, ethical and environmental cues. For example, a large number of Gen Z opts for vegetarianism and veganism on the basis of ethical reasons, to preserve the dignity of the animals (Bogueva & Marinova, 2020). In USA, 65% of Gen Z find plant-based foods appealing and 75% report to be reducing their meat consumption while in UK, 25% of Gen Z state that they have become vegan in the last year and 35% are actively seeking a "vegan partner" for dating (Schroeder, 2019). Understanding Gen Z's eating behaviours will not only shape future food preferences but will lead, support or even stop new technological developments and novel consumer products.

Investigating Food Behaviours

If we keep making food the way we do, we will also destroy the habitats of most wild plants and animals driving countless species to extinction. They are our life-supporting system; if we lose them, we will be lost too. Greta Thunberg (Climate change activist, Gen Z, born 2003)

Greta Thunberg's strong words are a damning assessment of human relationships with food. From subsistence agriculture to today's globalised industrial systems and the emerging novel food technologies, there have been incredible advancements that have changed the way foodstuffs are produced and how they reach the consumer's plate. However, the importance of food to provide nourishment, to comfort us and to play a primary role in connecting people is unlikely to change. In the 1970s, social sciences began to focus on the subject of food behaviour. The folklorist Don Yoder brought food under the wing of folklore studies (Long, 2009) and emphasised the need to investigate food behaviours popularising the term "foodways" to describe the range of habits, behaviours, customs and the entire gamut of cultural practices and beliefs surrounding eating. The French anthropologist and ethnologist Claude Lévi-Strauss (1979), in his book *The Raw and the Cooked* used an anthropological approach to study the symbolic use of food in human culture, tailored around the fundamental raw/cooked axis emphasising raw as of natural origin and cooked as cultural and product of human creation. Many years later in the 1980s and 1990s, researchers started applying the term "food studies" to investigate the historical, cultural, behavioural, biological and socio-economic determinants and consequences of food production and consumption. In contrast to Lévi-Strauss, the anthropologist Sidney Mintz (1985) produced a quintessential example of food studies research in his book *Sweetness and Power*, in which he traced the ways a single food substance, namely sugar, transformed modern history and culture.

Nowadays novel food technologies, including cultured meat, plant-sourced meat alternatives, algae- and insects-based products are having their turn in transforming the present and future food production systems and foodways. While the products we consume are poised to change, social scientists are dealing with food's influence on world events, distribution of power, global supply inequalities and social determinants of hunger as well as of diet-related conditions that cause non-communicable diseases, such as obesity, diabetes, heart disease and cancer. Nutritionists and public health experts are examining diets and eating disorders while psychologists are analysing food-related phobias and the connection between eating and taste, pleasure or disgust. The area of food research is burgeoning with different methodologies (Miller & Deutsch, 2009), historical perspectives (Belasco, 2002; Counihan et al., 2018) and futuristic approaches (Hamada et al., 2015; Levkoe et al., 2016) being applied.

Belasco (2008, p. x) succinctly summarises this thriving area by describing food behaviour as a "…complex negotiation among three competing considerations: the consumer's identity (social and personal), matters of convenience (price, skill, availability), and a sense of responsibility (an awareness of the consequences of what we eat)". This is the practical approach we also take to understand Gen Z's food behaviours. We will look at these three elements in a reverse order starting with responsibility.

Gen Z and Sense of Responsibility

For too long we have been waging a senseless and suicidal war on nature. António Guterres (Secretary-General of the United Nations, born 1949)

Despite the great interest in studying different food-related aspects, only in recent years have we started addressing the environmental consequences of our food consumption (Raphaely & Marinova, 2016; Bogueva et al., 2018). It is a complex relationship. The methods used to produce food, and livestock husbandry in particular, severely impact on the planet's ecosystems and these changes reflect human values. Beliefs, attitudes and values related to food, as represented in the trends in global production and consumption, are now seen as the main reason for the senseless and suicidal war on nature at the root of environmental degradation. Understanding the human dimensions related to food, including the human causes of, responsibility about and responses to environmental change, is essential to stop further degrada-tion and move towards regeneration of nature. The emerging and underexplored Gen Z cohort already plays a centre stage role in this. They are literally the new kids on the block.

Gen Z is not only different from all previous generations in terms of their food behaviours and daily habits, they are the cohort that surprises with its solutions and constantly challenges others. Compared to previous generations, they are not only digital natives – born in the era of digital technologies and the internet. They are digital integrators with technology fully integrated into their lives. Also digital translators, they know how to use the digital devices in functional and structural ways that deliver the best outcomes for them, including finding information, con-necting with others, sharing of ideas and mobilising for action (McCrindle, n.d.). They are pursuing causes that excite them, freely accept people as they are – racially and ethnically diverse, with different lifestyles and gender preferences. More self-aware and not as self-centred, willing to be a part of the solution for the world problems they encounter and are even willing to pay more for products that follow the same principles and share the same values as they do (Merriman, 2015).

As part of the 2021 Australian Science Week, we presented invited talks for schools in Perth and Sydney to a range of students – from the youngest Gen Z (Year 7) to the oldest at school (Year 11 and Year 12). What we noticed was sharp knowledge and values these young people have. This in no way should underesti-mate the wonderful job their teachers have been doing, but it was for the first time in our interactions with school-age children that we realised the power of this genera-tion. They could link the shrinking shape of glaciers to climate change, explain the weather patterns and had already designed their own sustainable and liveable city. When asked about which human activity is responsible for the largest use of land on this planet, a couple of girls unequivocally replied: "livestock". Gen Z already shows that they are on the right directional path when it comes to planetary health.

The same signs of generational transformation were revealed by older Gen Z 18–25 years old, who live in the major Australian cities of Sydney, Melbourne, Brisbane, Perth, Adelaide and Canberra and participated in our 2021 study. When asked about the main contributors to climate change, Gen Z generated a long list of issues, including coal mining, burning of fossil fuels, toxic gases, pollution, cars, buildings, manufacturing, transportation and power generation as well as food. Livestock and unsustainable agricultural practices were seen as contributing to cli-mate change by 45% of the 478 participants. The positive signs are there but Gen Z needs to do a lot more to take responsibility in the area of food.

Notwithstanding its information savviness, Gen Z does not seem to be fully aware of the fact that meat consumption has quadrupled globally in the last 50 years resulting in the world population consuming twice as much on a per capita basis than the previous generations. This trend has been growing exponentially. With our appetite, the war on nature has been affecting human health, animal welfare and areas of wilderness on this planet (Raphaely & Marinova, 2016; Bogueva et al., 2018; Willett et al., 2019; Morris et al., 2021). We are yet to witness a strong stance by Gen Z around the issue of using animals as food. However, among the surveyed

Gen Z (n = 478), more than a third, namely 38% (n = 182), believed that livestock production and animal-based food consumption are having a very significant contribution for lessening the well-being of the planet. Gen Z is already starting to realise that they need to also take responsibility for cleaning up the mess past generations have left them when it comes also to food. No other generation, even people who present themselves as environmentalists, has the same level of awareness about the ecological footprint of humanity's food choices.

Similar findings have emerged from other studies showing that the majority of Gen Z has limited awareness and understanding of the impacts of our food systems and those who are more knowledgeable rely on social media topics, most frequently related to plastics pollution and animal welfare (FSA, 2020). The previous generations are to a certain degree responsible by creating an inertia in building eating habits without questioning the status quo. Also, a large fraction of Gen Z is still living at home, where it has less control over its eating preferences (FSA, 2020). Nonetheless, the UK Gen Z is more likely to abstain or reduce its meat intake stating environmental reasons for such a choice (FSA, 2020).

Parallels can be drawn with the fossil-fuel based energy production and consumption where it took a lot of effort to build a momentum for change to renewable sources. Gen Z is protesting and engaging in action against fossil fuels. People like Greta Thunberg are also starting to unite Gen Z in taking responsibility for the food it eats. We are yet to see what this means as Gen Z grows older.

Gen Z and Convenience of Food Choices

My body will not be a tomb for other creatures. Leonardo Da Vinci (Italian painter, engineer, scientist, theorist, sculptor, architect and draughtsman, born 1452)

Despite the lack of deep knowledge and often unfamiliarity with the ingredients used by scientists and food producers to create novel food (Bogueva & Marinova, 2020), Gen Z instinctively feels that its relationship with food is governed by a technological, textural and flavour complexity combined with the magnitude of the surrounding environmental and sustainability context for our food systems and future food options. Gen Z is ready to change the world and the way we eat. These young people are there to monitor and make the food creation process personalised, helping to prevent diseases, demanding healthier choices and transparency in production (Bogueva & Marinova, 2020). They are ready to embrace the right diet needed in today's extreme climate emergency conditions. Is it only convenience that will determine this new food direction and are there any other critical factors? What will Gen Z want to eat?

Similar to other generations, food impacts Gen Z in many ways, making them emotionally and socially connected in their choices. Gen Z consumers want healthier, more convenient (Siegner, 2019), fresh and wholesome food (Campisi, 2020; Bogueva & Marinova, 2020). With their food preferences, it is not coming as a surprise that over the next five years Gen Z will drive a 55% growth in the fresh fruit category (NPD, 2019). Compared to the previous generations, Gen Z is raised with more education around the benefits of fresh vs processed food and with emphasis on food quality, which is evident in their very clear and overt preferences for the consumption of vegetables and fruits over other alternatives to meat (NPD, 2019; Bogueva & Marinova, 2020). These food preferences are likely to contribute towards a better life outlook (Mujcic & Oswald, 2016) of a generation already concerned about the natural environment and the world around them. Such food choices could be beneficial for their physical and mental well-being (Conner et al., 2017) as well as for planetary health.

Gen Z is about to make its mark related to alternative protein food choices. If we want to predict our future food, we need to understand this generation. The growing global population, the constraints on the Earth's natural resources, clearing of native vegetation, deforestation, antimicrobial resistance and consumer changes towards flexitarianism determine the need for value-added premium alternative protein sources to complement animal-derived food production. Whether Gen Z is prepared to embrace the new alternative protein choices and for what reasons – ethical, environmental or health related, is still unclear as they have a different agenda in mind. For example, a 2020 study shows that 72% of Gen Z are not ready to accept cultured meat as a novelty food and view it with disgust and fear (Bogueva & Marinova, 2020). Potentially eating cultured meat is also viewed as a betrayal to the national Australian identity of many Gen Z. If novel protein such as cultured meat is to replace livestock-based products and be widely accepted, it must emotionally and intellectually appeal to the Gen Z consumers (Bogueva & Marinova, 2020).

Nearly six in 10 participants in the Australian study (Bogueva & Marinova, 2020) expressed concern about the impacts of traditional livestock farming on the natural environment. Although 41% of the Gen Z study participants recognise the potential for cultured meat to become a viable nutritional source that is more sustainable and animal friendly than traditional meat, they require more transparency about commercial interests and potential future impacts. They also feel negatively about the involvement of large corporations and monopolies in the food sector (FSA, 2020). Many saw cultured meat and edible insects as a conspiracy and orchestrated agendas by the rich and powerful and were determined not to be convinced to consume these options (Sogari et al., 2019; Bogueva & Marinova, 2020).

Thinking about the future of food, Gen Z is highly worried about the environmental impact of its dietary choices and less about the global nature of the food system. Gen Z believes technology to be the key to delivering food quality, variety and value, presented with less harmful impacts (FSA, 2020) because this generation already has deeply formed perceptions and beliefs about anything related to being sustainable and eco-friendly. They have the same expectations related to food and are demanding authenticity, freshness and purity. While other generations, including Millennials and Gen X, are more ready to eat something mimicking meat or other plant-sourced alternatives, Gen Z prefers natural food sources especially real plants, fruit and vegetables (Food Frontier, 2019).

Gen Z and Food Identity

Eating for me is how you proclaim your beliefs three times a day. Natalie Portman (Academy award-winning Hollywood actress and activist for plant-based foods, born 1981)

The complexity with food choices is determined by many factors ranging from biological (hunger, appetite and taste), economic (cost, income and availability), physical ability and knowledge (education, cooking skills and time), social (culture, family, peers and meal trends), psychological (mood, stress, guilt, disgust and neophobia), attitudes and beliefs. Gen Z is not an exception in this complexity and does not fit into small, neat boxes (Deloitte, 2019). They are loud, visible, leaving a trace in the world. Their respect for diversity makes a clear mark not only with openness toward different cultures and respectful attitude toward others around them, but also with their food choices.

Growing up surrounded by endless and broad varieties of food from across the globe, Gen Z is now an irreversible part of the global foodie culture with a well-developed and an impressive for their age ample palate of tastes which differentiates them from any previous generations. They freely experiment and indulge without hesitation on different types of ethnic and international cuisine, from Asian (Indian, Thai, Japanese, Chinese, Vietnamese or Malaysian), to Middle Eastern (Lebanese, Iranian or Turkish), European (Italian, French, Greek, Portuguese, Spanish, Russian or Serbian), African (Moroccan or South African) and South American (Brazilian, Mexican or Cuban) enjoying the variety of flavours (Campisi, 2020). Different legumes, grains, vegetables, roots and nuts as well as new types of plants are all part of Gen Z's food culture.

The identity and emerging power of Gen Z are also identified as healthy eating trendsetters. Typically known for their preferences for fresh and wholesome foods, or at least one fresh component to be added to their everyday meals, Gen Z is turning its sight toward plant-based diets. An analysis of 3 million American students across 500 higher-education institutions shows that 65% are in favour of more vegetables in their diet and 79% are reducing their meat intake by adopting a flexitarian diet (Jed, 2018). Gen Z students are also reducing the use of single-use disposable plastics with plastic straws completely eliminated.

Respect for the natural environment and for animals transpire in Gen Z's food identity. These young people want their food to be part of the solution, not the problem (Abaño, 2020). Many are opting out from meat influenced compassionate and ethical reasons (Bogueva & Marinova, 2020). A study by the Australian Macadamia Industry reveals that about 42% of Australian Gen Z responded positively to the eating less meat trend while in China these people look for feel-good food (Wheeler, 2020). In USA, 44% of Gen Z believe that "being vegan is cooler than smoking" (Abaño, 2020, para. 2). A similar trend toward plant-based options, meaning vegetables and fruits, reports another survey of Gen Z (Deloitte, 2021). They are looking for quality cues, transparency, sustainability credentials, simple ingredients, nutrient density, new and exciting flavour experiences. Gen Z is ready to approach food with a sustainability-based determination, including plant-based options, mindful eating of meat, waste reduction and eco-packaging (Wheeler, 2020). A nationally representative panel survey of over 12,000 Australian adults showed that fruit and vegetable intake predicted increases in happiness, life satisfaction and well-being with improvements occurring within two years and such consumption might be a long-term investment in future health (Mujcic & Oswald, 2016). Gen Z is part of this trendsetting. Veganism is also the most popular nutrition topic on social media discussed by 54% of food influencers (Brandwatch, 2018); many of Gen Z are regularly following them and adopting vegan or vegetarian plant-based diets.

Gen Z's relationship with food is also shaped by the atmosphere and experiences they have when visiting restaurants. In 2018 in USA, Gen Z made 14.6 billion restaurant visits representing 25% percent of the total foodservice traffic (NPD, 2019). The cohort also identifies with organic and non-genetically modified food, beverages (Schroeder, 2019) and bowls, namely Acai bowls, burrito bowls or poke bowls. They demand authenticity, freshness, purity, flexibility, convenience, serving their personalised nutritional preferences and most importantly, the ability to link their identity with these trends.

With their experienced palate for diverse world tastes, Gen Z consumers are no strangers to novelty and they will open the floodgates of the food industry to flavour, texture, functionality and product innovation, but with an unlike approach. They are demanding difference in the future of food (Wheeler, 2020) and are as much as possible natural and solution-driven in their food choices. It is important to consider Gen Z cohort's eating habits not only in shaping our future food production, but for creating more resilient systems. They are the generation that will influence our future food choices where health, convenience and social media play important roles (Campisi, 2020). Gen Z wants convenient, healthy, local and organic food, with enhanced food experience, considering its moral and economic implications (Bumbac et al., 2020). It wants food options that are perceived as a positive choice (Wheeler, 2020). These young people are actively looking for greater control and transparency about the food they eat, including its origins, in order to make ethical and healthy choices (Bogueva & Marinova, 2020). Gen Z is not so concerned about their consumption behaviour, which could be transformed, but about the conduct and effects of foods inside their bodies – the benefits, negatives, the right healthy choices and avoidance of unhealthy nutrients. They prefer products and ingredients that do not create problems but are part of a solution, relevant to their lives and pertinent to what matters to them.

Many of the Gen Z are also quite confident in the kitchen, familiar and enjoying cooking at home (Hartman Group, 2018). Boosting their interest, creativity and confidence with booming cooking shows, such as MasterChef Junior TV competition, You Tube video tutorials or Instagram channels, Gen Z clearly wants to be in control and expects to be more involved in the preparation of their own food and meals. They want to experiment different options in the kitchen, controlling the flavour, taste of the final product or adjusting possible herbs as final touch (Bumbac et al., 2020). Gen Z's identity is already very much associated with plant-based food choices and these young people demonstrate their beliefs with every meal.

Gen Z and the Food Systems Industry and Policy

If I had an hour to solve a problem and my life depended on the solution, I would spend the first fifty-five minutes determining the proper question to ask, for once I know the proper question, I could solve the problem in less than five minutes. Albert Einstein (theoretical physicist, born 1879)

Transitioning to better food systems requires a dietary shift and Gen Z is well-positioned to be an active part in this transformation, particularly if given the right information and the opportunity to solve the problem. There are many recommendations and overarching strategies of how a food transition can be done. The list includes the EAT-Lancet Commission's Planetary Health diet (Willett et al., 2019), the Sustainable Healthy Diet Guiding Principles by the Food and Agriculture Organisation and the World Health Organisation (FAO & WHO, 2019), the Sustainable Agrifood Systems Strategies by the European Centre for Development Policy Management – an independent think-tank developing policies for inclusive and sustainable development in Europe and Africa (ECDPM, n.d.), and the European Union's Farm to Fork and Biodiversity Strategy, central to the European Green Deal (its new growth strategy for 2050) aiming to accelerate the transition to a sustainable, fair, healthy and environmentally friendly food systems (EC, 2020). These ambitious plans and targets require a large degree of involvement of the younger generation as active changemakers, and especially their role in the design of the solutions to improve the current food system. Through its behaviour and eating choices, Gen Z's role is pivotal in influencing trends and adopting new solutions, novel products, services, policies and strategies that help create a more resilient and just food system prepared for the future and responding to today's challenges.

There is no doubt Gen Z is projected to be a disruptor generation that will be having a profound, seismic impact on the food systems and the food industry globally. Having the potential to frame its food consumption choices in relation to climate, food waste (Kymäläinen et al., 2021) and animal welfare, to respond to health and wellness concerns, the expectations of Gen Z are to influence the use of the present and develop the future novel technologies and food innovations transforming and resetting production and industry priorities (Mintel Press Team, 2018). Prioritising convenience, functionality and added nutrients (NPD, 2019) combined with their expressed favouritism for fresh and wholesome food (Campisi, 2020) and creativity (Mintel Press Team, 2018), Gen Z can surely sway politicians and influence even dietary and nutritional guidelines. They expect from food companies and their research and development efforts to deliver foods that are produced in a transparent, engaging and conscientious way with environmental care and a clear label attached to them.

Gen Z's food preferences and attitudes will inform developments in many product categories and the outcomes will help shape practice and policy with benefits arising for the human society. Given the global food transition that has to occur in response to environmental threats and the planetary emergency, the actions of this new cohort as they relate to our future should be considered as something that is set to bring immense public gain.

Concluding Remarks

Young people – they care. They know that this is the world that they're going to grow up in, that they're going to spend the rest of their lives in... They actually believe that humanity, human species, has no right to destroy and despoil regardless. David Attenborough (natural historian, broadcaster and a cultural icon for biodiversity, born 1926)

In today's highly connected, especially for Gen Z, world, it is no surprise that similarities about changing expectations and food behaviours are now appearing in many countries around the world. Gen Z is pushing the need to change the conversation and learn from its own behaviours and attitudes. For these young people, eating is not based exclusively on factors such as the hedonic value of food, tradition, consumption habits, social status, but mostly on the perception, level of information and care.

We tried to paint a picture of Gen Z but there is a lot that the future will tell about this human cohort and what makes it tick. The desire for fresh and healthy food combined with transparency will continue to shape its eating behaviour. The world will need to embrace Gen Z's digital presence in the information space and attitudes for change, to cultivate its passion for urgent action to put a stop to practices that trigger climate change and reduce biodiversity. Food may soon become a uniting agenda. Having Gen Z change the world its way is maybe exactly what is needed to rejuvenate the planet.

One of Bella's challenges is: "Research sustainable food sources. What have you found?" (Burgemeister, 2019, p. 10). This is the last chapter of our book and we hope we have given you some answers. Continuing with Bella's (Burgemeister, 2019, p. 40) words: "I challenge you to make a difference! What will you do?"

References

Abaño, J. (2020). *'Feel-good food' is now a social currency for Gen Z*. Inside FMCG. https://insidefmcg.com.au/2020/10/16/feel-good-food-is-now-a-social-currency-for-gen-z/

Alter, C. (2018). *The school shooting generation has had enough*. Time Magazine. https://time.com/longform/never-again-movement/

Alter, C., Haynes, S., & Worland, J. (2019). *Time 2019 person of the year – Greta Thunberg*. https://time.com/person-of-the-year-2019-greta-thunberg/

Ayele, S., Khan, S., & Sumberg, J. (2017). Introduction: New perspectives on Africa's youth employment challenge. *Institute of Development Studies Bulletin, 48*, 1–12. https://doi.org/10.19088/1968-2017.123

Belasco, W. (2002). Food matters: Perspectives on an emerging field. In W. Belasco & P. Scranton (Eds.), *Food nations: Selling taste in consumer societies* (pp. 2–23). Routledge.

Belasco, W. (2008). *Food: The key concepts*. Berg Publishers.

Bogueva, D., & Marinova, D. (2020). Cultured meat and Australia's Generation Z. *Frontiers in Nutrition, 7,* 148. https://doi.org/10.3389/fnut.2020.00148

Bogueva, D., Marinova, D., & Raphaely, T. (Eds.). (2018). *Handbook of research on social marketing and its influence on animal origin food product consumption.* IGI Global.

Borg, R., & Aleah, K. (2020). *Year in review: How Black Lives Matter inspired a new generation of youth activists.* Rolling Stone. https://www.rollingstone.com/politics/politics-features/black-lives-matter-protests-new-generation-youth-activists-1099895/

Brandwatch. (2018). *Food influencers: The biggest food trends of 2018.* https://www.brandwatch.com/blog/react-food-trends-2018/

Bumbac, R., Bobe, M., Procopie, R., Pamfilie, R., Giuşcă, S., & Enache, C. (2020). How Zoomers' eating habits should be considered in shaping the food system for 2030 – A case study on the young generation from Romania. *Sustainability, 12,* 7390. https://www.mdpi.com/2071-1050/12/18/7390/pdf

Burgemeister, B. (2019). *Bella's challenge: A kid's take on the 17 UN Global Goals for Sustainable Development* (2nd ed.). The Book Incubator.

Campisi, V. (2020). *Gen Z's influential food preferences.* The Food Institute. https://foodinstitute.com/focus/gen-z-preferences/

Conner, T. S., Brookie, K. L., Carr, A. C., Mainvil, L. A., & Vissers, M. C. (2017). Let them eat fruit! The effect of fruit and vegetable consumption on psychological well-being in young adults: A randomized controlled trial. *Public Library of Science (PloS) One, 12,* e0171206. https://doi.org/10.1371/journal.pone.0171206

Counihan, C., Van Esterik, P., & Julier, E. (2018). *Food and culture: A reader* (4th ed.). Routledge.

Davey, M. (2016). *Australian students recreate Martin Shkreli price-hike drug in school lab.* The Guardian. https://www.theguardian.com/science/2016/dec/01/australian-students-recreate-martin-shkreli-price-hike-drug-in-school-lab#:~:text=A%20group%20of%20Australian%20high,hedge%20fund%20manager%20Martin%20Shkreli.&text=The%20drug%20is%20used%20to,infection%20in%20people%20with%20HIV

Deloitte. (2019). *Welcome to Generation Z.* https://www2.deloitte.com/content/dam/Deloitte/us/Documents/consumer-business/welcome-to-gen-z.pdf

Deloitte. (2021). *Global 2021 Millennials and Gen Z survey: A call for accountability and action.* https://www2.deloitte.com/content/dam/Deloitte/global/Documents/2021-deloitte-global-millennial-survey-report.pdf

Enterprise Nation. (2021). *The entrepreneur selling the benefits of the nut that's not a nut.* https://www.enterprisenation.com/learn-something/ayesha-grover-strpd/

European Centre for Development Policy Management (ECDPM). (n.d.). *Sustainable agrifood systems strategies.* https://ecdpm.org/dossiers/sustainable-agrifood-systems-strategies/

European Commission (EC). (2020). *Farm to fork strategy: For a fair, healthy and environmentally-friendly food system.* EU Green Deal. https://ec.europa.eu/food/system/files/2020-05/f2f_action-plan_2020_strategy-info_en.pdf

Finnegan, W. (2020). *Environmental activism goes digital in lockdown – But could it change the movement for good?* The Conversation. https://theconversation.com/environmental-activism-goes-digital-in-lockdown-but-could-it-change-the-movement-for-good-137203

Food and Agriculture Organisation of the United Nations (FAO), & World Health Organisation (WHO). (2019). *Sustainable healthy diets: Guiding principles.* Rome. http://www.fao.org/3/ca6640en/ca6640en.pdf

Food Frontier. (2019). Hungry for plant-based: Australian consumer insights. https://foodfrontier.org/wp-content/uploads/2019/10/Hungry-For-Plant-Based-Australian-Consumer-Insights-Oct-2019.pdf

Food Standards Agency (FSA). (2020). *The future consumers – Food and generation Z.* Food Standards Agency Research Report. UK. https://www.food.gov.uk/sites/default/files/media/document/generation-z-full-report-final.pdf

Fridays for Future. (2020). *Global digital strikes.* https://fridaysforfuture.org/next-big-strike-april-24/

Glover, D., & Sumberg, J. (2020). Youth and food systems transformation. *Frontiers in Sustainable Food Systems, 4*, 101. https://doi.org/10.3389/fsufs.2020.00101

Hamada, S., Wilk, R., Logan, A., Minard, S., & Trubek, A. (2015). The future of food studies. *Food, Culture & Society, 18*(1), 167–186. https://doi.org/10.2752/175174415X14101814953846

Hughes, E. (2021). *The age of entrepreneurism*. The Accountancy Partnership. https://www.theaccountancy.co.uk/business-start-ups/the-age-of-entrepreneurialism-78731.html

Hunjan, R. (2016). *Daraprim drug's key ingredient recreated by high school students in Sydney for just $20*. ABC News. https://www.abc.net.au/news/2016-11-30/daraprim-nsw-students-create-drug-martin-shkreli-sold/8078892

Ida, T. (2021). *Climate refugees – The world's forgotten victims*. World Economic Forum. https://www.weforum.org/agenda/2021/06/climate-refugees-the-world-s-forgotten-victims/

International Fund for Agricultural Development (IFAD). (2019). *Creating opportunities for rural youth: 2019 rural development report*. Rome. https://www.ifad.org/documents/38714170/41190221/RDR2019_Overview_e_W.pdf/699560f2-d02e-16b8-4281-596d4c9be25a

International Institute for Environment and Development (IIED). (2020). *Climate activism in the time of Covid-19*. https://www.iied.org/climate-activism-time-covid-19

Jed, E. (2018). *Aramark brings Gen Z food trends to life on college campuses nationwide*. Vending Times. https://www.vendingtimes.com/news/aramark-brings-gen-z-food-trends-to-life-on-college-campuses-nationwide/

Keneally, M. (2019). *11 deadly mass school shooting that happened since Columbine*. ABC News. https://abcnews.go.com/US/11-mass-deadly-school-shootings-happened-columbine/story?id=62494128

Kymäläinen, T., Seisto, A., & Malilia, R. (2021). Generation Z food waste, diet and consumption habits: A Finnish social design study with future consumers. *Sustainability, 13*, 2124. https://doi.org/10.3390/su13042124

Laybourn-Langton, L., Rankin, L., & Baxter, D. (2019). *This is a crisis: Facing up to the age of environmental breakdown*. IPPR. http://www.ippr.org/research/publications/age-of-environmental-breakdown

Lethlean, J. (2020). Thomas King, CEO of Food Frontier, 24. *The Weekend Australian Magazine*. https://www.theaustralian.com.au/weekend-australian-magazine/thomas-king-ceo-of-food-frontier-24/news-story/bb3a10d2aec68a8f1e24130c15636cc4

Lévi-Strauss, C. (1979). *The raw and the cooked* (J. Weightman & D. Weightman, Trans.). Octagon Books.

Levkoe, C. Z., Brady, J., & Anderson, C. R. (2016). Introduction. Toward an interdisciplinary food studies: Working the boundaries. In C. R. Anderson, J. Brady, & C. Z. Levkoe (Eds.), *Conversations in food studies* (pp. 3–22). University of Manitoba Press.

Long, L. (2009). Introduction. *The Journal of American Folklore, 122*(483), 3–10. https://doi.org/10.2307/20487643

Marinova, D., & Bogueva, D. (2019). Planetary health and reduction in meat consumption. *Sustainable Earth, 2*(3). https://doi.org/10.1186/s42055-019-0010-0

McCrindle. (n.d.). *How to speak Gen Z: The alphabet of Generation Z on flip cards*. Available at: https://mccrindle.com.au/insights/blogarchive/how-to-speak-gen-z-the-alphabet-of-generation-z-on-flip-cards-infographic/

Merriman, M. (2015). *What if the next big disruptor isn't a what but a who? Gen Z is connected, informed and ready for business*. Ernst & Young. https://www.ey.com/Publication/vwLUAssets/EY-what-if-the-next-bigdisruptor-isnt-a-what-but-a-who/$File/EY-what-if-the-next-big-disruptor-isnt-a-what-but-a-who.pdf

Miller, J., & Deutsch, J. (2009). *Food studies: An introduction to research methods*. Berg Publishers.

Mintel Press Team. (2018). *IFT18: Generation Z set to impact the future of food and drink innovation*. https://www.mintel.com/press-centre/food-and-drink/generation-z-set-to-impact-the-future-of-food-and-drink-innovation

Mintz, S. W. (1985). *Sweetness and power: The place of sugar in modern history*. Viking.

Morris, C., Kaljonen, M., Aavik, K., Balázs, B., Cole, M., Coles, B., … White, R. (2021). Priorities for social science and humanities research on the challenges of moving beyond animal-based food systems. *Humanity Social Science Communication, 8*, 38. https://doi.org/10.1057/s41599-021-00714-z

Mujcic, R., & Oswald, A. J. (2016). Evolution of well-being and happiness after increases in consumption of fruit and vegetables. *American Journal of Public Health, 106*, 1504–1510. https://doi.org/10.2105/AJPH.2016.303260

NPD. (2019). *Gen Zs are getting older and making their mark on restaurants and eating trends*. The NPD Group. https://www.npd.com/news/press-releases/2019/gen-zs-are-getting-older-and-making-their-mark-on-restaurants-and-eating-trends/

NPR Staff. (2019). *Transcript: Greta Thunberg's speech at the U.N. Climate Action Summit*. https://www.npr.org/2019/09/23/763452863/transcript-greta-thunbergs-speech-at-the-u-n-climate-action-summit

Oatsu. (2021). *Oatsu overnight oats*. https://oatsu.co.uk/

Overdeep, M. (2020). *Texas teen helps save honeybees with wildly successful lemonade business*. Southern Living. https://www.southernliving.com/news/me-and-the-bees-lemonade-teen-ceo-mikaila-ulmer

Phospholutions. (2021). *Developing solutions to increase the efficiency of global phosphorus use*. https://www.phospholutions.com/

Raphaely, T., & Marinova, D. (Eds.). (2016). *Impact of meat consumption on health and environmental sustainability*. IGI Global.

Reid, K. (2021). *Forced to flee: Top countries refugees are coming from*. World Vision. https://www.worldvision.org/refugees-news-stories/forced-to-flee-top-countries-refugees-coming-from

Reserve Bank of Australia (RBA). (n.d.). *The global financial crisis*. https://www.rba.gov.au/education/resources/explainers/the-global-financial-crisis.html

SBS News. (2021). *'I fear for my Afghan sisters': Malala Yousafzai sounds urgent warning following Taliban takeover*. https://www.sbs.com.au/news/i-fear-for-my-afghan-sisters-malala-yousafzai-sounds-urgent-warning-following-taliban-takeover?dlb=[2021/08/18]%20del_newsam_bau_02&did=DM2524&cid=sbsnews:edm:acnewsam:relation:news:na:na

Schroeder, B. (2019). *How Generation Z is creating the opportunity of a lifetime. Pay attention as this is not a fad but a deep long-lasting trend*. Forbes. https://www.forbes.com/sites/bernhardschroeder/2019/09/13/how-generation-z-is-creating-the-opportunity-of-a-lifetime-pay-attention-as-this-is-not-a-fad-but-a-deep-long-lasting-trend/?sh=3460cfd42bf8

Seemiller, C., & Grace, M. (2017). Generation Z: Educating and engaging the next generation of students. *Practice, 22*(3), 21–26. https://doi.org/10.1002/abc.21293

Seemiller, C., & Grace, M. (2019). *Generation Z: A century in the making*. Routledge.

Siegner, C. (2019). *Gen Z consumers want healthier, more convenient food*. Food Dive. https://www.fooddive.com/news/gen-z-consumers-want-healthier-more-convenient-food/548299/

Slezak, M., & Timms, P. (2021). *Australian teenagers' climate change class action case opens 'big crack in the wall', expert says*. ABC News. https://www.abc.net.au/news/2021-05-27/climate-class-action-teenagers-vickery-coal-mine-legal-precedent/100169398

Sogari, G., Bogueva, D., & Marinova, D. (2019). Australian consumers' response to insects as food. *Agriculture, 9*(5), 108. https://doi.org/10.3390/agriculture9050108

Storch, L. V., Ley, L., & Sun, J. (2021). New climate change activism: Before and after the Covid-19 pandemic. *Social Anthropology, 29*(1), 205–209. https://doi.org/10.1111/1469-8676.13005

strp'd. (n.d.). *Guilt-free baking with tigernut flour*. https://www.strpd.com/

The Hartman Group. (2018). *Gen Z 2018: Today's teens – Tomorrow's adults*. https://www.hartman-group.com/reports/2106932950/gen-z-2018-todays-teens-tomorrows-adults

U Report. (n.d.). *U-Report climate change bot*. https://ureport.in/story/947/

United Nations (UN). (2020). *The climate crisis – A race we can win. Shaping our future together*. https://www.un.org/en/un75/climate-crisis-race-we-can-win

United Nations Children's Fund (UNICEF). (n.d.). *Toolkit for young climate activists*. https://www.unicef.org/lac/en/toolkit-young-climate-activists

United Nations Educational, Scientific and Cultural Organisation (UNESCO). (2019). *Behind the numbers: Ending school violence and bullying.* https://unesdoc.unesco.org/ark:/48223/pf0000366483

United Nations Environment Program (UNEP). (2021). *Young people call for urgent climate action at UN environment assembly.* https://www.unep.org/news-and-stories/story/young-people-call-urgent-climate-action-un-environment-assembly

United Nations High Commissioner for Refugees (UNHCR). (2021). *The refugee brief.* https://www.unhcr.org/refugeebrief/latest-issues/

Valuing Children Initiative. (2021). *Bella Burgemeister.* http://valuingchildreninitiative.com.au/ambassador/bella-burgemeister/

Vow. (2020). *We make really, really good food.* https://www.vowfood.com/what-we-do

Wheeler, M. (2020). *Generation Z demanding 'different' in the future of food.* Macadamia. https://foodmag.com.au/tag/macadamia/

Willett, W., Rockström, J., Loken, B., Springmann, M., Lang, T., Vermeulen, S., … Murray, C. J. L. (2019). Food in the Anthropocene: The EAT–Lancet Commission on healthy diets from sustainable food systems. *The Lancet, 393*(10170), 447–492. https://doi.org/10.1016/S0140-6736(18)31788-4

Workman, J. (2019). *Our house is on fire.* 16 year-old Greta Thunberg wants action. https://www.weforum.org/agenda/2019/01/our-house-is-on-fire-16-year-old-greta-thunberg-speaks-truth-to-power/

World Health Organisation (WHO). (2021). *Suicide.* https://www.who.int/news-room/fact-sheets/detail/suicide

Yami, M., Feleke, S., Abdoulaye, T., Alene, A. D., Bamba, Z., & Manyong, V. (2019). African rural youth engagement in agribusiness: Achievements, limitations, and lessons. *Sustainability, 11*, 185. https://doi.org/10.3390/su11010185

Young, E. (2021). *Thousands join climate strikes across Australia, angry with the government's gas investments.* SBS News. https://www.sbs.com.au/news/thousands-join-climate-strikes-across-australia-angry-with-the-government-s-gas-investments

Epilogue

It is only with the heart that one can see rightly; what is essential is invisible to the eye.
Antoine de Saint-Exupéry, *The Little Prince*

Food in a planetary emergency is not just an environmental and climate change problem, it is also a social question. Human and industry behaviours have slowly changed the face of our planet and now we need to deal with these consequences. Climate change, biodiversity loss, alterations of the nitrogen and phosphorous cycles, land- and sea-use changes are upon us, they are all happening and are human induced. It is no longer a matter of avoiding risks but responding to certainties. We have the tools and capacity to understand the changes but do we have the ability to minimise and wherever possible, revert the impacts?

At the end of the twentieth century, Vitousek (1994) called for those who understand the ecological changes to get active, get connected, get real and don't get down, as individually and collectively it was possible to make a difference. Thirty years later, we are left with a much smaller window of opportunity to take action and a lot of the changes have become irreversible. However, we want to believe that human society individually and collectively has advanced in understanding the dramatic nature and causes of the planetary emergency and is now prepared to act.

This book presented a lot of material with scientific facts and insights but there are many other perspectives and issues we did not cover. For example, does the food of our close-to-a-billion human pets need to also change? We only briefly touched on issues related to the changing nutritional values of different foods but that aspect requires a deeper analysis and understanding. What is the place of genetically modified (GM) species and GM food in the planetary emergency? Investment trends and the destiny of start-up companies in food technologies surrounding the 7th technological wave also deserve to be followed. Do food distribution channels need to change and what is the role of different digital apps linking producers and consumers? As with the electricity transition from fossil fuels to renewable sources, is there a role for prosumers? Animal welfare and the importance of re-wildering the planet are similarly not discussed. Are we heading towards an irreversible disaster or

D. Marinova, D. Bogueva, *Food in a Planetary Emergency*,
https://doi.org/10.1007/978-981-16-7707-6

should we have faith in humanity, and particularly the younger generations? Should the response to the climate emergency be left as a question of self-discipline or should it be imposed on people? This list can be further expanded…

Food is a basic requirement for survival but also has so many tacit aspects that you can only feel from the heart (de Saint-Exupéry, 2010). The beat of 7.9 billion human hearts will give the answer to the current and future generations about what is essential and how we see their future.

References

De Saint-Exupéry, A. (2010). *The little prince*. Penguin Press.
Vitousek, P. M. (1994). Beyond global warming: Ecology and global change. *Ecology, 75*(7), 1861–1876. https://doi.org/10.2307/1941591

Ingram Content Group UK Ltd.
Milton Keynes UK
UKHW020146090523
421397UK00002B/8